Chicago School of Data

Chicago School of Data

A regional ecosystem in the service of people

THE SMART CHICAGO COLLABORATIVE

edited by Denise Linn Riedl

To the people who do the work.

The gross national product does not allow for the health of our children, the quality of their education or the joy of their play. It does not include the beauty of our poetry or the strength of our marriages, the intelligence of our public debate or the integrity of our public officials. It measures neither our wit nor our courage, neither our wisdom nor our learning, neither our compassion nor our devotion to our country, it measures everything in short, except that which makes life worthwhile. And it can tell us everything about America except why we are proud that we are Americans.

— Robert F. Kennedy, Remarks at the University of Kansas, March 18, 1968

How can I assemble data that will increase the caring quotient in our community?

— Terry Mazany, Remarks at Chicago School of Data Days, 2014

"Data! data! data!" he cried impatiently. "I can't make bricks without clay."

— Sir Arthur Conan Doyle, The Adventure of the Copper Beeches

Manufactured in the United States of America by the
Smart Chicago Collaborative
http://www.smartchicagocollaborative.org / @smartchicago
UI Labs
1415 N. Cherry Ave.
Chicago, IL 60642
(773) 960-6045

Supported by the John D. and Catherine T. MacArthur Foundation.

Set in Scala and ScalaSans

Library of Congress Control Number: 2015953051
ISBN: 978-0-9907752-3-2
First Printing, 2017

Contents

Introduction

Written by Daniel X. O'Neil, former Executive Director of the Smart Chicago Collaborative

> *"The Smart Chicago Collaborative is all about collaboration, working to define, introduce and organize, bring together, entities—the people, tools, organizations, institutions, processes and policies—that are in this ecosystem of data and to create definition to that ecosystem. Why does this all matter? It matters because the problems we face are daunting, the consequences of failure are devastating, and time to act is short. That means if you can do it by yourself, it probably isn't worth doing."*
>
> —TERRY MAZANY, CEO & PRESIDENT, THE CHICAGO COMMUNITY TRUST, WELCOMING REMARKS ON SEPTEMBER 20, 2014

The Chicago School of Data—or, simply, "the ecosystem project"—was born out of the decades-long work of the The John D. and Catherine T. MacArthur Foundation in funding and shepherding data intermediaries for Chicago nonprofits.

The discipline of using data to make lives better in Chicago goes back at least as far as Jane Addams and her work mapping tuberculosis outbreaks. More recently, the Metro Chicago Information Center, which existed from 1990 to 2012, served as a central place for neighborhood groups, nonprofits, and other institutions to go to for classic data intermediary work. These functions—holding and describing data, interpreting data for constituents, performing technical work on datasets—have now been split among a number of organizations in the region.

During this same period, there has been an increase in the number and sophistication of players in the space. A lot of this work is centered around the University of Chicago, some can be traced back to the focus on data in the Obama presidential campaign, and the Emanuel administration has pushed forward lots of data generation and analysis efforts. Great work has come out of places like DePaul University, Woodstock Institute, and LISC Chicago. Smart Chicago has also emerged as an important and learned worker in the space.

Then there's the vast number of organizations that use data to do their jobs—whether they feed the hungry, provide beds for the homeless, bring arts and culture for the masses, and so on. With months of outreach, we were able to pull together a unique and deep grouping of great workers.

In short, there has been an abundance of effort, an eruption of growth, an increase in funded projects, but a paucity of alignment in the sphere of using data to serve people in Chicago. This project seeks to change that.

This "Chicago School"

Chicago has a long tradition of schools of thought supported by leading intellectual institutions, such as the Chicago School of Economics, the Chicago School of Architecture, and the Chicago School of Sociology.

The Chicago School of Data is a thoughtful and practical movement focused on the connection between people and data in Chicago. We spent the time making connections with people across our region to determine their relationship to data. Our goal is to connect practitioners in our space and develop a collaborative framework for improving these connections across the Chicago data ecosystem.

We deviated from the traditional school of thought because we wanted to include everyone. We wanted to reach any and all organizations that use data in the service of people despite the type of data they collect, the tools they use, or the skills they have in using data. We knew that this project would only be of value if it was inclusive and exhaustive.

Components of this Work

There are three main components associated with this project: a scan of the field, documentation and mapping of the landscape, and a conference to convene the workers in this space.

Scan of the Field

We wanted to convene and sharpen the focus of a core group of practitioners in Chicago who use data to improve the lives of residents. This built on the existing work of the "Assessment of the Community Information Infrastructure in the Chicago Metropolitan Area" from the National Neighborhood Indicators Project and other convenings. We assembled a core stakeholders group comprising the City of Chicago, Cook County, MacArthur Foundation, and LISC Chicago to advise us and guide our work.

We did an immense amount of outreach to more than 1,000 organizations via phone calls and emails. We received census forms from 258 people from 236 different organizations. We conducted nearly 90 in-depth interviews. By listening to organizations, we began to understand roles, connections, dependencies, and potential collaborations between organizations in the Chicago data ecosystem. We were also able to identify and discuss opportunities to bridge gaps.

What we heard from organizations drove our 2014 conference, Chicago School of Data Days— a two-day experience wholly based on the feedback we have received from these surveys, months of interviews, and listening to people at work.

Documentation and Mapping of the Landscape

The second part of this project was to map what we learned about the data work happening in Chicago—the entities, companies, enterprises, civil service organizations, and other groups that make up the field. We want to create a cohesive narrative around this landscape that gives shape, direction, and clarity to everyone included.

This book will be the main deliverable of this component.

Through the duration of this project we shared interviews and analysis. Here is a piece from Andrew Seeder, a key project team member, who began to document and classify this data landscape in 2014:

> *"After months of interviews and hundreds of surveys we're beginning to see how the regional data ecosystem fits together. The ecosystem grows and develops because we create data for others to use, we consume data made by others, and we enable each other to do the same. We found data creators, data consumers, and data enablers.*
>
> *Some organizations create packaged data sets of data they've collected, while other organizations make it a business of cleaning free, public data. Others donate hardware and their expertise to local schools or, as an institution, they fund organizations working in the field. But data creators consume data and data consumers enable others to create data. These broad categories aren't mutually exclusive."*

Chicago School of Data Days

At the start of this project, the Chicago School of Data Days conference was meant to be a time and place to come together and share our findings and discuss what the ecosystem is. As we did the work, we learned that the conference was a bigger and more important opportunity to convene people who may never have been in the same room together. As we were listening to practitioners who worked with myriad tools, processes, and methods, Chicago School of Data Days became a conference about sharing experiences, talking about resources, and meeting and learning from one another.

As such, our sessions were based on surveys and interviews. Our speakers were people who we interviewed, and our audience became what we referred to as the "fourth speaker," who shared about their own use of data. Almost 300 people came to the conference, and we documented each session with notes, livestreams, videos, photographs, and tweets to guide this book.

In This Book

We were surprised by the number of organizations who already saw themselves as part of the data ecosystem. The people we spoke with understood the importance of this work and that data can further their organization's mission:

> *"We very much understand the need for comprehensive data, both to manage our current business and to help forecast into the future. Data is a key piece, which then comes alive in the narrative about the clients we serve."*
>
> —SOL FLORES, FOUNDING EXECUTIVE DIRECTOR, LA CASA NORTE

We defined major themes that we heard in the surveys and interviews, these themes informed our conference agenda: Gaps, Skills, Tools, Sharing & Privacy, Accessing Data, and On-Ramps.

In this book, we cover in detail what we learned about Chicago's current data ecosystem and our process to get to this point. We cover details about outreach, interviews, documentation and conference logistics. We will describe the roles each project team member played leading up to the conference, and the process of gathering information to do the ecosystem analysis. This book is our attempt to map the data landscape and share processes on this particular project in the hopes that our work can be helpful to others.

References

http://www.smartchicagocollaborative.org/toward-a-structure-for-classifying-a-data-ecosystem/

http://www.neighborhoodindicators.org/library/catalog/assessment-community-information-infrastructure-chicago-metropolitan-area

Terry Mazany, then CEO of the Chicago Community Trust, addresses the Chicago School of Data Days participants on September 20, 2014 *(Photo by Daniel X O'Neil)*

Participating Organizations

The Chicago School of Data was built to be inclusive. We are not just data collectors or advanced or sophisticated data consumers. We cared about everyone, so when it came time to organize the Chicago School of Data Days, we invited everybody.

Below is the full list of participants in our scan of the field and the Chicago School of Data Days:

#33cc77:

741 Collaborative

Access Community Health Network

Active Transportation Alliance

Adler Planetarium

After School Matters

AIDS Foundation of Chicago

Albany Park Theater Project

Alliance for Illinois Manufacturing/NORBIC

Alphonsus Academy and Center for the Arts

American Red Cross

Andersonville Chamber of Commerce

Archdiocese of Chicago

ARkay Solutions

ArtReach at Lillstreet

Arts Alliance Illinois

Association House of Chicago

Back of the Yards Neighborhood Council

Baxley's Village

Bethel New Life

Big Shoulders Fund

Bottom Line

Breakthrough

Bridge Communities

BUILD

Catalyst Group Global

Center on Wrongful Convictions

Chicago Federation of Labor Workers Assistance Committee

CHANGE Illinois

Changing Worlds

Chapin Hall at the University of Chicago

Chatham Business Association, SBDI

Chicago Appleseed Fund for Justice

Chicago Architecture Foundation

Chicago Arts Partnerships in Education

Chicago Botanic Garden

Chicago Cares

Chicago Children's Museum

Chicago City Data Users Group

Chicago Commons

Chicago Community Data Project

Chicago Cook Workforce Partnership

Chicago Federation of Labor Workers Assistance Committee

Chicago Heights Veterans Center, Department of Veteran Affairs

Chicago Jazz Philharmonic

Chicago Jobs Council

Chicago Justice Project

Chicago LGBT Homeless Youth Task Force

Chicago Lights Tutoring and Summer Day

Chicago Public Library

Chicago Public Libraries Archer Heights Branch

Chicago Public Library Foundation

Chicago Public Schools

Chicago Run

Chicago Sinfonietta

Chicago Teachers Union

ChildServ

Christopher House

Citizen Advocacy Center

Citizen Schools

City of Chicago

City Year

Civic ArtWorks

Co-Knowledge

Sarah Macaraeg (Columbia College Chicago and independent projects)

Communications, Languages and Culture, Inc

Community Media Workshop

Council for Adult and Experiential Learning

CR Threads LLC

Crain's Chicago Business

Creative Partners

CREED Consulting

Crown Family Philanthropies

Data Science for Social Good

Data Science for Social Good Fellowship

DataMade

Datascope Analytics

Deborah's Place

Delta Institute

DePaul University: The Red Line Project

Doejo

DonorFuse

DonorPath

Donors Forum

DuPage Children's Museum

DuPage Federation on Human Services and Reform

Lola Chen (East Garfield Park advocate)

Education Systems Center at Northern Illinois University

Emphanos

Enlace Chicago

Family Focus, Inc.

Family Resource Center on Disabilities

Family Shelter Service

First Folio Theatre

Foresight Design Initiative

Foundations of Music

Free Spirit Media

FUSE

Gary Comer Youth Center

Get IN Chicago

Golden Apple Foundation

Greater Auburn Gresham Development Corporation

Hadiya's Promise

Halcyon Theatre

Harvard University

Have Dreams

Healthy Schools Campaign

HHCS

Housing Options for the Mentally Ill

Hoyne Associates, Inc.

IBM

Illinois Campaign for Political Reform

Illinois Institute of Technology: Boeing Scholars Academy

Illinois Legal Aid Online

Illinois Mentoring Partnership

Illinois Sentencing Policy Advisory Council

Impact Engine

Katya Lysander (independent data consultant)

Ingenuity

Institute for Housing Studies

Institute for Justice Clinic on Entrepreneurship

Jane Addams Resource Corporation

Joyce Foundation

Kartemquin Films

Kelly Hall YMCA

Krontiris Niemczewski

La Casa Norte

LAF

Lakeview Pantry

Lawyers' Committee for Better Housing

Leyden Family Service and Mental Health Center

LISC Chicago

Literacy Works

Loaves and Fishes Community Services

Logan Square Neighborhood Association

Lumity

Media Burn Independent Video Archive

Mercy Housing Lakefront

Metropolitan Planning Council

Microsoft

Midwest Pesticide Action Center

Mikva Challenge

Metropolitan Planning Council

Museum of Contemporary Art Chicago

Museum of Science and Industry Chicago

Namaste Charter School

National Hellenic Museum

National Latino Education Institute

Neighborhood Housing Services of Chicago

Network for College Success

Network for Teaching Entrepreneurship

New Life Centers of Chicagoland

North Lawndale Employment Network

Northwest Side Housing Center

Northwestern Memorial Hospital

OAI, Inc.

Oak Park-River Forest Community Foundation:

Oak Park River Forest Food Pantry

Office of Mayor Rahm Emanuel

One Million Degrees

Onward Neighborhood House

Openlands

OrangeBoy, Inc.

Partnership for a Connected Illinois

Peggy Notebaert Nature Museum

PODER

PositivEnergy Practice

Private

Project Exploration

Project Tech Teens

Public Good Software

Puerto Rican Cultural Center

Respond Now

Restoration Ministries, Inc.

Rogers Park Business Alliance

Safer Foundation

SBS Computer Center

Kristi Leach (self)

SGA Youth and Family Services

Shimer College

Skill Scout

Smart Museum of Art

Social IMPACT Research Center at Heartland Alliance

Socrata

South Asian American Policy and Research Institute

South Suburban Mayors and Managers Association

St. Agatha Family Empowerment

St. Pius V Church

Stern Consulting

Streetsblog Chicago

Strengthening Chicago's Youth

Su Casa Catholic Worker

Symbol Training Institute

Technology Access Television

Kobie Robinson (representing a technology start-up)

The Ark of St. Sabina

The Cara Program

The Chicago Public Education Fund

The Children's Place Association

The CivicLab

The Resurrection Project

Tutor/Mentor Institute, LLC

United Way of Metropolitan Chicago

Unity Park Advisory Council

Adrian Ciccone (University of Chicago)

University of Chicago Consortium on Chicago School Research

University of Chicago Medicine Urban Health Initiative

University of Illinois - Chicago

UNO Charter School Network

Urban Gateways

Urban Initiatives

We the People Media/Residents' Journal

West Humboldt Park Development Council

Windy City Habitat for Humanity

Women Employed

Woodstock Institute

World Business Chicago

YMCA of Metropolitan Chicago

YMCA of the USA

Young Chicago Authors

Youth Outreach Services

Youth Service Project

Zealous Good

Making sense of our data ecosystem meant understanding the common themes surrounding data challenges, gaps, strengths, and areas for potential collaboration in the city. There was a good reason that the Chicago School of Data Days were not organized around organizations' types (such as consumers of data, collectors of data, analysts, advocates, trainers)—namely, that the shared challenges and goals of mission-driven data users ended up being more important than the roles they had or the types of institutions where they worked. As a result, these shared challenges became the center of gravity around which we built the Chicago School of Data Days and this book.

The raw responses from the Chicago School of Data participants are public. Those results are summarized broadly in our Current State of the Ecosystem chapter, as well as broken down by theme in the next several chapters: Gaps, Sharing & Privacy, Skills, Accessing Data, On-Ramps, and Tools. See the Meta chapter of this to understand our methods for outreach that helped us achieve a comprehensive, inclusive scan of our participants.

References

http://www.smartchicagocollaborative.org/a-taxonomy-for-regional-data-ecosystems/

http://www.smartchicagocollaborative.org/toward-a-structure-for-classifying-a-data-ecosystem/

https://gist.github.com/danxoneil/c21d85f96c3b5abc85a9

https://docs.google.com/spreadsheets/d/1ALP5vZCwkf6hNn8BH_UNY-3IwDxHTeCAm7JAWVTPyy20/edit#gid=0

Gaps

"There would be a huge benefit to nonprofit and social service agencies sharing data because there are a lot of organizations doing the same work. There is no way for one organization to know what another organization is doing because we are so siloed. Everybody is holding really tight to their information, and doesn't want to share, so even if we cross that huge hurdle of getting tools, tech, and training in the hands of the organization ... how do we get over that siloed attitude?"

—PARTICIPANT AT CHICAGO SCHOOL OF DATA DAYS, INFRASTRUCTURE SESSION

Despite the challenges to using data, it seems like everyone agrees that data is important. Among different kinds of organizations, each with its own mission, there's little agreement about why data is important, how to get it, use it, and what to do with it. Phrases like "data-driven" and "results-based" are used as proof that an organization uses data to achieve its mission or operate efficiently.

In this Gaps chapter we will take inventory and organize the Chicago organizations' challenges to meaningful data use, as seen in the Chicago School of Data survey and the discussions at the Chicago School of Data Days. We will discuss how affordability, organizational capacity, and access to data itself can limit how well organizations can do this work.

Here's what members of the Chicago School of Data thought were the greatest challenges to working with data:

- 141 practitioners said that they are unable to dedicate the time to work with data given other demands
- 110 practitioners said staff lack the necessary technical skills to work with data

- 79 practitioners are unable to gain access to the data they need
- 69 practitioners said they are unable to afford the tools necessary to make use of data

Organizations experience gaps in capacity, affordability of certain data tools and expertise, and access to data. Beyond the survey results measuring the state of the whole ecosystem, we wanted to highlight important organizational cases surrounding data infrastructure and capacity in organizations, affordability gaps, and access gaps. During the conference, we gave practitioners a space to articulate the limits they come up against in the field and share tips about how to overcome those limits.

Gaps in Infrastructure & Capacity for Data Use

The first panel addressing data gaps at the Chicago School of Data Days was on "Infrastructure," or, the internal capacity of organizations undertaking data work. The role of collecting, analyzing, and using data falls under so many different job roles, and are sometimes only a small piece of a person's job at an organization. Through our interviews, too, we heard again and again that organizations were unable to dedicate time to work with data, and they believed that their staff did not have the technical skills needed to work with data. We realized that few organizations have a staff position that solely focuses on data, and that there is a desire and a need to use data better throughout organizations.

Understanding How Data Can Drive Mission

Margaux Pagan, then Managing Director of DonorFuse, recommended that organizations go back to basics and think about reasons why they want to use data in the first place. They should think about storytelling and shaping numbers with words. They should think about how data will support their mission and how they can leverage data to make clear choices that make an impact. Pagan emphasized that data "silos" should be broken down — that organizations should work in the open and, in general, be more aware of how data is shared internally and with partners.

Building an Internal Culture for Data

In the last few years, LISC Chicago has given a lot of thought to its data culture, and over time more data has been collected and used for decision-making. Taryn Roch, Program Officer of Evaluation & Impact, shared that back in 2012 it was important to simply assess LISC's capacity for collecting and using data. Support, resources, and manpower were added, but it was not completely a smooth transition. In the words of Roch, there were a few complicating factors:

> *"Neighborhood boundaries are porous, so how do you measure where people come from? How do you decide on a time horizon for an evaluation? How do you develop internal capacity to address data needs?"*

Since LISC works on many collaborative projects across the city, Roch's perspective on data and organizational change was also formed by what she observed from partners. Roch explained that, in general, organizations were empowered to address barriers in ways that fit their needs. For example, at the Chicago Lawn Housing Initiative, training existing staff (one on one) and hiring new data coordinators were absolutely crucial steps. But more than just having the people and skills, it was important to have vision. That took strong leadership, and a sense of how data capacity fits into the larger framework of the mission.

Roch provided two takeaways during her talk at Data Days:

1. Realize that causation is not always clear or possible to prove
2. Enable reflection and encourage learning within the organization

On the theme of increasing organizational capacity for data use, Jill Young, now Senior Director of Research and Evaluation at After

School Matters, also stressed the importance of leadership around data.

Young discussed how a staff position around research and evaluation was added to After School Matters to focus on outcomes and indicators. A culture shift happened. Asking, "What is your impact?" became important for the data team. With support from the board and chief program officer, the team developed a common language around data, put a logic model in place as a roadmap for growth, and created key partnerships with Chicago Public Schools to access data, all of which moved everyone forward.

Affordability Gaps

The second type of gap addressed through the Chicago School of Data Days was the affordability gap that exists across institutions in Chicago working with data. Panelists were Spencer Cowan, formerly of the Woodstock Institute, Stephen Pigozzi of the Association House in Humboldt Park, and Samia Malik of the Chatham Business Association. Each provided a different perspective on affordability challenges.

A Community Center's Perspective on the Price of Data Management

Association House is a long-standing settlement house in Humboldt Park providing workforce development and digital skills training. Like other community centers and training facilities, Association House has funders that require some form of reporting. Stephen Pigozzi, the AmeriCorps & Technology Center Supervisor for Association House at the time of the Chicago School of Data Days, shared a common challenge: funders expect results and proof of impact, but funders might not be willing to invest in the work or tools needed to sustain data tracking.

A Data Intermediary's Perspective on the Price of Accessing Data

Woodstock provides research, data analysis, and technical assistance to different organizations across the city. They classify themselves as a data intermediary—instead of working directly with residents, they work with the organizations that work directly with residents.

"Affordability gaps are relative," Spencer Cowan pointed out. He explained that his organization works with and secures public or affordable data. For Woodstock, $6,000 meant affordable. Cowan acknowledged that it might not be affordable to other organizations with different budgets or data priorities.

In its data intermediary role, Woodstock can speak to two types of affordability gaps:

1. The price of accessing high-quality data, which Woodstock experiences as an organization
2. The price of providing technical assistance to mission-driven community organizations, which Woodstock absorbs

"You'd be surprised what we can do in four hours." Cowan said. He pointed out that a community organization equipped with the right data or map for their cause can be essential.

A Business Association's Perspective on the Price of Data Gathering

The price and energy associated with data gathering for the Chatham Business Association stemmed from technology gaps prevalent in the community:

- 80% of the businesses they work with don't have a website
- 35% don't have email
- 45% don't have Internet access at their businesses

Without email addresses, they could not contact businesses. Without internet connections, how would the business fill out forms and input their data? To address the technology divide that impacted the quality of their data collection, Chatham Business Association created the Get Connected program.

Samia Malik, a Project Manager at the Chatham Business Association, talked about one of the biggest problems that they face:

not having "an online footprint." Given the constraints presented by this technology gap, Chatham Business Association goes door-to-door, conducting surveys to collect data. Fortunately, the strong relationships they have with local businesses give them a higher response rate to the surveys. Unfortunately, they lose out on a lot of data from South Side and West Side communities. Also, the data sets that they receive are not always accurate.

Affording the Tools and Software Your Organization Needs

There is a price to securing the software and tools to meet your organization's data needs. This price is both in time and money.

At the time of the Data Days Conference, the Chatham Business Association had secured their first ArcGIS license—a tool that made them optimistic for future work. However, learning the program takes time, and they will probably only use 10% of the software's capabilities.

Pigozzi narrated the annual battle in which he negotiates to keep an imperfect data management system for Association House, ETO (Efforts to Outcome).This story sparked an interesting suggestion about how smaller organizations in Chicago can avoid such situations.

One participant in the session suggested that the Chicago Benchmark Collaborative jointly purchase software. He also mentioned the possibility of building a custom, modular, data system for community centers. Another audience member suggested that big software companies waive their licensing fees for products that are "overbuilt" for small organizations.

See the Tools chapter later in this book for a list of recommended open source or discounted tools.

Data Access Gaps

As the Chicago School of Data evolves, the accessibility of reliable data remains a challenge to its growth. Kathy Pettit of the Urban Institute began the conference by mentioning that looking for data often feels like "looking for a needle in a haystack." Later in the conference, Terry Mazany, then President of the Chicago Community Trust asked thought-provoking questions about data access and equity of information: "Who has access to these data and who does not? Are we increasing disparities or using data as a force for good to reduce disparities?"

In the School of Data Survey, we heard that organizations are not sure how to access some of the data they need. The Access Gaps session at the Chicago School of Data Days featured speakers with stories about barriers to accessing data, where organizations find the data, and how organizations work together to share data or data systems.

Collaborative Model can Create Meaningful Data Across Nonprofits

In the session on Access Gaps, Traci Stanley, the Director of Quality Assurance for Christopher House, spoke of her involvement in the Chicago Benchmarking Collaborative: "We were all tracking outcomes of our programming, but we were getting questions from our boards about how we compare to similar social service agencies." In the nonprofit world, benchmarks don't really exist, and if they do, "you feel like you are comparing apples to oranges," said Stanley.

Initially a group of five, the Chicago Benchmarking Collaborative "came together for comparative insights" and to improve the quality of data on nonprofit outcomes in Chicago. How do you compare programs and target populations, so you can know that you are comparing apples to apples?

Now a group seven agencies, the Collaborative engaged and outcomes expert and purchased Efforts to Outcomes (ETO) software to build their own reports and track outcomes and create consistency in the data. ETO "is really flexible and worked for a number of different programs." The cross-agency data reporting created greater accountability; "it has helped identify effective program strategies." Programming changes are now driven by data results.

Stanley's presentation sparked several questions from the audience on funding increases from the project. She answered, "Funders really embrace the data ... Since we are all competing for funding, it took a lot of trust for us to work together."

Access to Data is not the Same Thing as Access to Their Meaning

With the Smart Chicago Collaborative, Tracy Siska, the Executive Director of the Chicago Justice Project, created a project called Crime and Punishment in Chicago. The Chicago Justice Project has also been focused on building a systems approach to data around sexual assault. They created a task force to determine how cases drop out of the system.

In Chicago from 2005 to 2009, there were 6,000 calls for service related to rape per year, but only 1400 reports per year, then 1300 and then 1200. While reports were declining, the number of calls for service were the same. But people began to incorrectly report that rape was declining in Chicago.

This is why Siska advocates for a systems approach to data as opposed to an incidence approach. "Using only incident data without a systems approach means that what makes it into the news is just wrong," said Siska. "The CPD is really good at capturing data, but not good at using it."

In conclusion, Siska recommended: "Do data about trends. If we don't know the trend, how do we know what a large increase is?"

Data Available on Schools and Students in Chicago

Eliza Moeller, then Director of the Data-Practice Collaborative at the University of Chicago, spoke about data and data products available through UChicago Impact and Chicago Public Schools (CPS).

"CPS has excellent data," Moeller said. CPS created the "freshmen on-track indicator" based on determinants of high-school success. The CPS Performance Website does a yearly school evaluation report. They make a large amount of data available and very often the data are broken down by school. There are still gaps, however. CPS lacks data on charter schools.

Moeller works with data to create useful reports. These reports are currently available at ccsr.UChicago.edu. Current reports include data on national freshmen on-track rates compared to the CPS average. There is also a report on projected college enrollment and college enrollment.

When discussing next steps for this project and these data, Stanley stated: "The goal is to move to an online format and really interact with it. That will come out through UChicago Impact."

References

http://www.smartchicagocollaborative.org/access-gaps-session-at-chicago-school-of-data-days/

http://www.smartchicagocollaborative.org/results-from-eliminate-the-digital-divide-advisory-committee-capstone-project/

http://consortium.uchicago.edu/

http://crime-punishment.smartchicagoapps.org/

https://docs.google.com/document/d/1eNZVv-qeF8IzosSkgP7JI90-VgfDLKnhITsgDXBXTuI/edit

https://docs.google.com/document/d/1pSqXIkim-8Pnbeet-vvEut3Y5s0X-4w5hBXS6bvS50_Q/edit

https://docs.google.com/document/d/17HsSGHcWf2vVyE_qCD3EgO-cW2xxWxk3KQNV7bgsqmv0/edit

Tracy Siska of the Chicago Justice Project showcases the website Crime and Punishment in Chicago during the "Access" session of the Chicago School of Data Days *(Photo by Carley Mostar, Chicago School of Data Documenter)*

Sharing and Privacy

"Sharing is trust. Privacy is power."
—MELISSA PIERCE, DIRECTOR OF CWDEVS

Several responses from the Chicago School of Data Census highlighted common difficulties that arise around accessing sensitive or proprietary data. We asked, "Is there data that you want to use but you can't because you can't get permission to use it? If so, what is it?" Responses included:

- "Health or education data with identifiers restricted due to HIPAA or privacy concerns"
- "Other organizations' data; it's a privacy/confidentiality issue"
- "Some data is student-level data, which is privacy protected"

The "Sharing & Privacy" sessions at the Chicago School of Data Days focused on how data may be shared responsibly, how to keep people safe when their information gets used, and what can be reasonably assumed to constitute an informed consent. In this chapter we present the key recommendations and themes from those sessions.

Data Sharing

Speakers Andre Kellum, former Executive Director of the 741 Collaborative, Kathryn Bocanegra, former Violence Prevention Director of Enlace Chicago, and Nate Inglis Steinfeld, the Research Director of Illinois Sentencing Policy Advisory Council, grappled with thematic questions surrounding data sharing: How can we create a culture of sharing across government and private organizations? What is the expected value of data sharing? Is sharing a core value that we think should exist throughout Chicago?

A Call to Break Down Data Silos

Steinfeld spoke about the siloed data in The Illinois Sentencing Policy Advisory Council (SPAC). This is how they describe their work:

SPAC was created to collect, analyze and present data from all relevant sources to more accurately determine the consequences of sentencing policy decisions and to review the effectiveness and efficiency of current sentencing policies and practices. SPAC reports directly to the Governor and the General Assembly. See 730 ILCS 5/5-8-8(f)

SPAC shares average offender profiles, proposed legislations costs, and trend analyses. At the time of the Chicago School of Data Days, SPAC was looking to connect data across subject areas, the ultimate goal being to create a cost-benefit model. That cost-benefit model would uncover the value of investing in social programs (e.g., early learning) and how those would affect the justice system.

Solving complex problems involves linking data across subject areas, sectors, and parts of government. The main take-away: don't assume that your data—whatever it is—isn't relevant to criminal justice research.

"I want to make a pitch to you all," Steinfeld challenged at the Data Sharing session at the Chicago School of Data Days. "Publish your information, and we'll see what we can do to link the data."

Models for Sharing Across Organizations

Katheryn Bocanegra, former Violence Prevention Director of Enlace Chicago, shared the story of her organization's quest to use data ethically and create a culture of data sharing.

Enlace Chicago is dedicated to making a positive difference in the lives of the residents of the Little Village community by fostering a physically safe and healthy environment in which to live and by championing opportunities for educational advancement and economic development. According to Bocanegra, Little Village has become a laboratory for experiments in data-driven policing and community development. The National Institute of Justice conducted the Gang Violence Reduction Project there in 2003; the University of Illinois, Urbana-Champaign studied the effects of crime on children's physical activity in 2011.

> *"Part of the process of creating a culture of community data-sharing has been to form shared metrics to measure kids' relative health: connection to caring adults, future aspirations, and attitude towards interpersonal peer violence. Our goal is to get kids out of the survival game, into a thriving game. It's not 'If I live til I'm eighteen,' but 'When I reach eighteen, this is what I'm gonna do with my life.'"*
> — KATHERYN BOCANEGRA

Enlace borrowed CPS's early warning indicators for defining at-risk youth, tracking factors such as failing a reading or math course, missing 20+ days of school, or behavioral incidents. They found that there were between 640-800 at risk youth that were in 5th through 8th grades. As of 2014, they were engaging 500 youth in various projects, and were collecting data in order to measure the long-term, longitudinal impacts of their work on community safety.

By sharing information on youth welfare, progress, and struggles, collaborators can better strategize to help youth in the neighborhood. Bocanegra related her struggle to get similar data from the 10th district police, a concession that took two years to wrangle, due to privacy laws regarding youth involved in violent crime. She is now able to track juvenile crime perpetration and victimization.

To create a culture of data sharing, Katheryn Bocanegra made these recommendations for organizations

- Choose shared metrics — it's a challenge, but a necessity
- Vet the database with community stakeholders
- Establish confidentiality measures
- Training, training, training ("On a weekly basis")
- Learn from the challenges

Enlace Chicago also created a trauma inventory, measuring individual kids' exposure to violence. "Hurt people hurt people," Bocanegra reminded her audience. "If I've seen my best friend shot, if I witness domestic violence at home, and then someone at school rubs me the wrong way, I'm much more likely to respond with aggression."

Enlace set up firm ethical and legal boundaries as well, establishing confidentiality measures and limiting access to the information. There are some vulnerable populations—particularly domestic violence survivors—about which organizations cannot share information, even with the confidentiality measures. Safety and trust have to be paramount in the community.

Another model for data sharing explored at the Chicago School of Data Days was the 741 Collaborative. The 741 Collaborative works with community members and community-based organizations to share data for the benefit of 4 Chicago neighborhoods: Douglas, North Kenwood, Grand Boulevard, and Oakland.

741 stands for 7 organizations, 4 communities, and 1 common goal. To make data sharing work, the collaborative brought in an outside facilitator to help develop opportunities for the partner organizations to improve. The facilitator also helped the organizations decide which organization did what best. 741 also created a part-time data position to work between the partner organizations. The value of this work wasn't in another shared database. The value was in individual organizations' reports, resources, and analyses—not just individual-level data. According to former Executive Director Andre Kellum, sharing data in this way makes organizations more efficient. More importantly, sharing data can help communities.

Privacy

Privacy is crucial to the strength of Chicagoland's data ecosystem. At Data Days, Matthew Bruce of the Chicago Workforce Funders Alliance, Vivian Hessel of the Legal Assistance Foundation for Metropolitan Chicago (LAF), and Matthew Roberts of the Chicago

Department of Public Health discussed how privacy concerns are addressed in their work. They also discussed how datasets can be prepared to respect people's privacy and protect against data breaches.

What is Responsible Data Sharing?

Bruce, Executive Director of the Chicago Workforce Funders Alliance, described how addressing privacy early in a data sharing collaboration helps bring best practices to the workforce development sector. These collaborations depend on sharing personal information to coordinate a job placement or develop a job training program for a neighborhood. Collaborations are high-stakes, as they demand that people's identities be kept private.

Matthew Bruce raised four key questions that need to be decided to responsibly share data: Who needs to know what and when? What are the objectives of sharing data? What does a release of information really mean? Where does liability ultimately lie?

Hessel, Director of Technology for Advocates at LAF, articulated similar questions addressing the technical challenges of using a dataset with personally identifiable information. Hessel pointed out that data can be identifiable even though it isn't thought of or even characterized as personal identifiable information. For example, if there is a dataset of employees at a medium-sized company that includes gender and age, it could be easy to deduce identities.

Recommended privacy questions to ask about personally identifiable data

- How sensitive is it? The more sensitive, the more safeguards needed.
- Whose data is it? If someone is trusting you with their data, you may need to take steps to protect it before you share it. Get their permission, remove personally identifiable data.
- What are the risks? If the risks are small, then sharing is easier.

- What are the responsibilities? If you have a responsibility to keep the data safe, take steps to fulfill it before you share.
- Who owns the data after you put it online? Are you giving up ownership? Will ownership change?
- Who can access the data? Is it encrypted? Are passwords required?
- How is the data stored?
- How is the data deleted? Is it truly deleted?

Balancing Privacy & Open Data in Government

Matthew Roberts, Informatics and Health IT Director of the Chicago Department of Public Health, emphasized that there is a balance between confidentiality and usefulness when it comes to data—especially health data. A health agency might be disincentivized from releasing data by confusing privacy laws, a lack of internal capacity to clean and analyze data, or a worry about the public misinterpreting the data. Despite those threats, Robert pointed out that released data can create unpredictable public value. For example, New York released bed availability data in nursing homes before Hurricane Irene. This inventory eventually helped get residents out of harm's way.

Informed Consent

The Chicago School of Data Days hosted a discussion on "informed consent," the process and ethics around asking permission before data is collected. David Eads, Melissa Pierce, and Matt Gee facilitated the group conversation about these challenges. The conversation also covered Institutional Review Boards (IRBs), sensors, and other surveillance mechanisms that spur questions concerning data ethics.

Definitions of Consent

To express how important informed consent is in the age of big data, Pierce, Director of CWDevs, used the language of sexual consent to frame the conversation about data collection: "Yes means yes. Consent means consent...We need to be clear. Yes equals yes."

For Pierce, informed consent around data is like the mutual consent of sexual relationships, something which involves real people's lives and their right to their own bodies. She explained that people take informed consent seriously when they see their data as an extension of themselves, a part of their body and their thoughts.

Gee pointed out that the past can help us answer questions surrounding definitions of consent. In August 1947, judges issued a verdict against Karl Brandy and 22 other Nazi doctors, whose medical regime sterilized 3.5 million German citizens, and who had themselves experimented on (tortured) people in concentration camps, ostensibly for the purposes of advancing "medical science." Part of the Nuremberg Trials, this verdict set the groundwork for the Nuremberg Code, 10 principles for ethical medical research.

10 Principles of the Nuremberg Code

1. Required is the voluntary, well-informed, understanding consent of the human subject in a full legal capacity.
2. The experiment should aim at positive results for society that cannot be procured in some other way.
3. It should be based on previous knowledge (like, an expectation derived from animal experiments) that justifies the experiment.
4. The experiment should be set up in a way that avoids unnecessary physical and mental suffering and injuries.
5. It should not be conducted when there is any reason to believe that it implies a risk of death or disabling injury.
6. The risks of the experiment should be in proportion to (that is, not exceed) the expected humanitarian benefits.
7. Preparations and facilities must be provided that adequately protect the subjects against the experiment's risks.

8. The staff who conduct or take part in the experiment must be fully trained and scientifically qualified.
9. The human subjects must be free to immediately quit the experiment at any point when they feel physically or mentally unable to go on.
10. Likewise, the medical staff must stop the experiment at any point when they observe that continuation would be dangerous.

Decades after the Nuremberg Code, three core virtues for medical research emerged in the Belmont Report (1978): Respect for persons, beneficence, and justice. Gee pointed out that some web-based technologies operate as a "non-consensual experiment" and said, "People who haven't thought about ethical experiments run them all the time." When personal data get used in large-scale web experiments, how are technology companies held accountable to these core virtues?

The concern about informed consent is due in part to uncertainties over how personal data will be used in the future. "There's no going back," said Eads. "What are the kinds of social contracts we need? How do we talk about this stuff? What are things going to look like in 40 or 50 years?"

User Agreements & Limitations

User agreements are a recognizable form of user consent. Data Day participants talked specifically about Google Glass, which Pierce was wearing at the time. They discussed whether it was possible to give informed consent to be recorded by a Google Glass device when you could be recorded by a Glass whenever you're near one—there's no way to even tell if the gadget is on or off. In that case, a user agreement may have applied to the person who bought the device, but not to all of the other people who indirectly interacted with it.

Another case discussed was the iTunes drop, which downloaded U2's album "Songs of Innocence" into every Apple iTunes subscriber's library. The song was framed as a "gift," not an invasion of privacy. Apple did something with their technology that some users weren't expecting, but was covered under Apple's user agreement.

References

Data Sharing notes https://docs.google.com/document/d/1ILfupqt_FoKjHQl6u4Cz-kBudKXqTCoh6m88ym3YDgI/edit

Data Sharing video https://www.youtube.com/watch?v=QusxX-CQ-7Kw&feature=youtu.be

Privacy notes https://docs.google.com/document/d/1wa_LDe2O1h8-byHm573obEWA2sY-NhbpYOe7MU_vwho/edit

Privacy video https://www.youtube.com/watch?v=_Y-mR2XWE9w&feature=youtu.be

Informed Consent notes https://docs.google.com/document/d/10-qbW-r3maEReALvimamLhgvWjjnBZtbW9PMMS_n8sxE/edit

Informed Consent video https://www.youtube.com/watch?v=-eoe6KVKy-qU&feature=youtu.be

Every tab Melissa Pierce (panelist in Informed Consent) had opened on her computer to get ready for this way too short conversation

http://papers.ssrn.com/sol3/papers.cfm?abstract_id=1926431

http://blogs.hbr.org/2013/04/the-hidden-biases-in-big-data/

http://www.katecrawford.net/pubs.html

http://mashable.com/2011/02/03/permission-marketing-social-data/

http://indieboxproject.org/blog/2014/09/lets-create-the-internet-of-our-own-things/

Matt Bruce of the Chicago Workforce Funders Alliance and Matthew Roberts of the Chicago Department of Public Health share their experiences at the "Privacy" session of the Chicago School of Data Days *(Photo by Nourhy Beatriz, Chicago School of Data Documenter)*

Skills

Through the Chicago School of Data survey, we took inventory of organizations' in-house skill sets. We asked organizations what they needed help with, and these were the responses:

- Basic computer literacy (18 organizations)
- Basic data literacy (68)
- Basic spreadsheet skills (39)
- Basic data analysis skills (81)
- Advanced data analysis skills (157)
- Data cleaning and preparation skills (103)
- Data management, storage, and retrieval skills (117)
- Data visualization and communication skills (150)
- Other skills (27)

Only a handful of organizations said they needed basic computer literacy. Deeper into the results we find that 10 of the 18 organizations who said they needed basic computer skills also said they needed help developing every other skill we listed. This aspect of the survey tells us that the ecosystem needs to accommodate organizations who want to develop basic computer skills and also know something about advanced data analysis.

We also asked, "Is there data that you want to use but can't because it's too hard to work with? If so, what is it?" Common responses pointed to CPS data, the City of Chicago Open Data Portal, or Census Bureau data.

We organized the Chicago School of Data Days Skills sessions so they spoke to the commonalities we saw in the survey responses: interests in diving deeper in data visualization, census data, and open source tools. This section shares the discussions, cases, and lessons that came out of those sessions.

Open Source

This session answered, "How do open source software projects work and how can organizations use them to get things done?" It covered an introduction to the fundamentals of GitHub, how to buy and maintain URLs, and how hosting works. Dan Sinker, the Director of Knight-Mozilla OpenNews and Dan O'Neil, former Executive Director of the Smart Chicago Collaborative, led this session.

The first task was to define open source. The session attendees landed on, "software with source code that is out there for anyone to look at," but Sinker pointed out that "open source" now means a lot more than that. Open source means that there is a license that allows for making a copy of the code, manipulating it on your own, and running it.

Four components of open source projects

- Open to inspect
- Able to run
- Available to change
- Possible to change

GitHub is the largest version-control software service, where open source projects are shared and forked. The set of norms governing version control software facilitates effective collaboration and avoids problems commonly found in collaborative document-making—think: files named "final," "final final," and "no, really, final."

An open source community requires great documentation, governance of code base, and a community of building and advocacy. Most importantly, open means being welcoming, sharing, and nurturing.

Open Source Beyond Code

Sinker pointed out that open source is no longer just code—it extends to hardware, furniture, books, and recipes. For example,

Sinker created tacofancy on GitHub. It is a repository of taco recipes which grew to over 200 recipes with the help of 75 contributors. Some contributors corrected spelling, others standardized formatting, someone wrote an index generator. People learned GitHub just to post their recipes. It was created in plain text and had a low barrier of entry, but was still very much an open source project.

"GitHub is a language you have to understand, but tacos are a much easier language to understand," said Sinker.

O'Neil pointed out that the Smart Chicago Collaborative itself strives to be an open source organization by the way it operates. "In life we're accepting pull requests all the time. It's being responsive to criticism," he said, comparing Smart Chicago's process to the way GitHub operates. Smart Chicago does things publicly and has collaborators on every project.

"I want to think about and talk about how we can apply the principles of open source to the offline work we do together," said O'Neil.

Data Visualization

During Data Days, we heard from Beckie Stocchetti, then the Community Engagement Manager at Kartemquin Films, Emily Withrow, Assistant Professor at Northwestern University, and Chris Hagan, Web Producer and Data Reporter for WBEZ. They shared applications that can help organizations through data visualization from start to finish.

Recommended Tools

There were several tools recommended by the Data Visualization Panel. OpenRefine, an open source preparation tool, helps you merge, match, de-duplicate, and clean data. Shan Carter's Mr. Data Converter converts data between different formats. For flat value delimited files, Google's Fusion Tables integrates data with other Google products and allows for the easy creation of charts and maps. Another mapping solution, QGIS, is a free and open geographic information system. Leaflet is an open source JavaScript library for people who want to make interactive web maps. GitHub's Open Journalism repository collection and NPR's explanation of "How to Setup Your Mac to Develop News Applications Like We Do" breaks down how journalists can create visualizations step by step.

Stochetti's 10 quick thoughts about data

- Be succinct! Distill data down.
- It's ok to show people what they already know.
- Data visualization can be static.
- Know when it's a good tool and when it's not. Be discerning.
- Think about how you'll organize data before you create surveys. Why are you collecting this info? I.e., when creating evaluation forms, think about how you will use the info you are collecting. Think about how to reduce information so that it's simple and understandable.
- Don't always collect more data than you need.
- Use the easiest data aggregator for your purpose.
- Don't disregard simple tools. Google Docs may be migrated into Google Graphs, for example.
- Learn to interweave data with a narrative. How do you use stats in a conversation?
- Expand the concept of data to include story.

The group pointed to the many free visualization suites online such as Quandl, easelly, or infogra.am. With some Google-fu you can find tutorials for specific software. For social media, Stocchetti and her team use Hootsuite to manage the content of all their profiles. Skills for social media are essential when an organization wants to gain wider exposure or develop its brand. Social impact and efforts-to-outcome analysis is key to successful data visualization, especially when an organization's audience is its board or a grantmaker.

See the Resources chapter of this book to see a full list of data visualization resources shared during the Chicago School of Data Days.

Census

Joe Germuska, Chief Nerd at Northwestern University's Knight Lab, led the Chicago School of Data Day's session on census data. The session focused on navigating U.S. Census data both through apps developed by the federal government as well as the homegrown Census Reporter tool.

The Census Reporter simplifies finding and using data from both the decennial census and the American Community Survey, and it offers data by geographic location and general topic.

The application has a friendly user interface with responsive visualizations. Germuska, project lead for Census Reporter, plus his team, share news of the application's success stories online. The team opened their source code to fetch files from the U.S. Census's FTP (file transfer protocol) interface. The Census hosts its data products in a tiered file structure, which they then serve to users through FTP. The Census Reporter's open source code makes working with Census data easier, and it enables others to work with the data without having to write a program themselves.

Other tools for using and analyzing census data:

- Census.IRE.org
- AmericanFactFinder
- IPUMS.org
- NHGIS.org
- Social Explorer
- Data Ferrett

For more information about census data sources and Census Reporter, watch Joe Germuska's presentation at Data Days.

References

Open Source Session Video https://www.youtube.com/watch?v=lZhrH8lp6wc&feature=youtu.be

Open Source Session Notes https://docs.google.com/document/d/1DBKsHfF2orWOQf-j2Wfpco3n8iWnwfHGsBy2ZDmM6bo/edit

Data Visualization Session Video http://youtu.be/tHSoCKw2d3w

Data Visualization Session Notes https://docs.google.com/document/d/1fP11tAvYWTP_rsSt48kovNY4LZfv9MC95XfdpvyC3Lo/edit

Census Data Session Video https://www.youtube.com/watch?v=LECREydWa9I&feature=youtu.be

Census Data Session Notes https://docs.google.com/document/d/1IuAk-geZMgvGnkM5EoDTqQmSzdZgUNiRP6CeSjhg88PQ/edit?usp=sharing

Brainstorming notes from the Chicago School of Data Days *(Photo by Julie Torkelson, Chicago School of Data Documenter)*

Accessing Data

Sometimes accessing data is an organization's biggest barrier to successfully using data. In this chapter we'll see how organizations overcome access barriers. We'll also cover different ways of accessing data, including online searching, regional data portals, formal data acquisition templates, and scraping web content.

79 members of the School said they couldn't access the data they need. The organizations they came from were both big and small, from direct service providers to research institutions. When asked, "Is there data that you want to use but you can't because you can't get permission to use it? If so, what is it?" some of the responses were:

- Many datasets owned by USDA are under confidentiality agreements
- Although the Circuit Court of Cook County has court case data, it is accessible online one case at a time. It would be nice to have a regular feed of data. The Electronic Docket Search interface is provided by Lexus, but not sure who to talk to about it
- CPS report card and standardized testing data
- Other organizations' data; it's a privacy/confidentiality issue data on CCC [City Colleges of Chicago] students from 4-year institutions—would require multiple data sharing agreements

Common themes included accessing Chicago Public Schools' data, data on youth, and data on health. One organization said: "We'd like to be able to keep scraping data that pertains to neighborhood issues—to give nonprofits (and journalists) context for what money is being spent in Chicago." It's also important to note that the challenges that organizations shared about accessing data generally often overlapped with other categories of conversation at the Chicago School of Data Days — especially privacy and affordability.

The Chicago School of Data Days organized a session around the data access challenges: data acquisition procedures and sharing agreements, leveraging regional data portals, and searching and scraping for data.

Data Acquisition

At Data Days, Sarah Duda, the Associate Director of the Institute for Housing Studies at DePaul University (IHS), and Susan Yanun, the former Director of Evaluation and Accountability at the Logan Square Neighborhood Association, spoke about how their organizations acquire and manage data. This session covered memorandums of understanding (MOUs) and data partnerships, among other things.

Data Sources & Sharing at IHS

Duda works at IHS, which transforms raw data into actionable information. IHS' mission is to provide reliable, impartial, and timely data and research to inform housing policy decisions and discussions in the Chicago region and nationally. They use data collection and cleaning, research, and technical assistance to inform housing policy.

At IHS they've created an easy-to-use clearinghouse for the region's housing data. The clearinghouse functions on top of several Memorandums of Understanding (MOU). MOUs are one way that two or more parties decide how data can be shared. The terms of the agreement change depending on circumstance. Institutional review boards (IRBs) are another way to guarantee that sensitive data is passed between people. Public documents can be accessed after completing a Freedom of Information Act request (FOIA). See Chapter 7, Sharing and Privacy to learn more about MOUs.

Core data sources of the IHS include the Cook County Assessor, the Cook County Recorder of Deeds, and the Cook County Clerk of the Court. Through these sources, IHS developed 16 indicators about housing market conditions, which includes composition of the housing stock, characteristics of sales, mortgage activity, foreclosure filings and auctions, and long-term vacancy. Stakeholders and vendors find value in the data IHS has acquired and repackaged. IHS's work helps them understand collection channels and other housing market issues.

IHS's data are granular, timely, flexible, and publically available. These strengths are not without their challenges, though. The data are designed for program administration, not analysis. The data require extensive development and expertise for interpretation.

One of the core challenges faced by IHS and many other organizations is how to make data useful for others. While IHS is a critical part of Chicagoland's data ecosystem, especially in terms of housing data, its primary audience is policymakers and other researchers. We also heard from another organization which uses data to evaluate its own programming, so that it may better serve neighborhood residents.

Acquiring Data From Parents & Students

The Logan Square Neighborhood Association's Parent Mentors program has removed barriers between school and home for many Logan Square young people, and it has demonstrated how parents can work together to improve the community. The program collects data to evaluate their success.

The data was helpful in identifying how LSNA could improve its program and envision where to go next. LSNA developed a Parent Engagement Institute to help parents understand what is happening in the classroom and, in turn, what impact the classroom is having on community outcomes. At the time of the conference, the next step was to formally evaluate its impact data.

LSNA collects data in several forms:

- Parent mentor pre-post surveys to gauge involvement in their children's school
- Teacher pre-post surveys to try and understand what's happening in the classroom
- Principal pre-post surveys

From these data, LSNA found that there's the most opportunity to train parents in specific areas. Then LSNA worked with consultants to identify what curriculum could best meet the needs of all of these parent-mentor situations. Based on this information, LSNA developed nine training modules.

Lessons learned from LSNA's surveying

- Devote resources (time and money) to data acquisition, troubleshooting, follow-up, and analysis
- Be as clear as possible with what it is you want to know
- Get buy-in on why results will be helpful
- Get input from the "experts" (such as principals/teachers)
- Check and double-check whether you need a consent form and if it contains what you need

There were questions Yanun mentioned that were of interest to LSNA, but which the data did not yet illuminate: What's within the parent-mentor sphere of influence—in what ways do they influence academic achievement? What do we know about the growth of students that work with parent mentors? What are strong indicators of academic achievement?

Both the IHS and the LSNA show how organizations can acquire data in different ways. IHS gets data through MOUs and then cleans the combined data into a public-facing clearinghouse. The data is especially useful to housing market researchers and analysts. The LSNA collects survey data about its parent mentoring program so it can understand how successful the program is and where it's having the most impact.

Regional Data Portals

Chicagoland's data ecosystem thrives on its regional data portals. At Data Days, representatives from different levels of government came together to discuss open data available online to nonprofits, small businesses, and residents. Simona Rollinson of Cook County, Derrick Thomas of Cook County, and Tom Schenk of the City of Chicago participated in the Regional Data Portal Session. Audience members learned about the types of datasets already available and how to find what they were looking for.

Cook County Open Data

"Open data is gaining momentum," said Simona Rollinson, Chief Information Officer of Cook County. A 2011 ordinance made Open GIS data available to the public and available for commercial, non-commercial, charitable, and educational purposes. The data is a result of a collaboration with Smart Chicago, without which Rollinson said they wouldn't be as far along as they are.

At the time of the Chicago School of Data Days conference, the most-accessed Cook County datasets were...

- Cook County Employee Annual Salaries back to 2011
- Awarded Contracts
- Cook County Foreclosures
- Check Register
- Quit Claim Deeds
- Map showing all Cook County Facilities and Service Locations
- Map of the Cook County Commissioner District
- Map with the GIS Address Points for Chicago
- Map with the GIS Address Points for Suburban Cook County

Derrick Thomas, Director of Application Development & Management for Cook County Government, introduced the data portal. While for many years the state denied FOIAs on GIS requests, that data is now available for things like a virtual cemetery run through the Medical Examiner's office. "It's very challenging to mine data across so many platforms," Thomas said. He stressed the importance of modernization, as different offices sit on different platforms. "If it's on the mainframe, I have to ask a programmer to write code to access it." Thomas said that "momentum is there" and they're taking steps, but "it hasn't happened yet."

The City of Chicago Open Data Portal

The City of Chicago's Tom Schenk, Chief Data Officer for the City of Chicago, took the audience on a tour through Chicago's data portal. He prefaced his tour by saying it had been the top-down push from Mayor Emanuel that spurred this work, and that the data availability became less about performance metrics versus helping out nonprofits and small businesses.

Schenk brought up the city's crime database, which started in 2001. It reports crimes that happened up to a week ago, and runs once a day. It displays the where and what, a location according to latitude and longitude, but of course, not who. Schenk said this data is often used for academic purposes or by the Chicago Tribune.

He moved onto another data set, highlighting the fact that Chicago is "the first government to publish energy data per building per block." He called the beach-quality data, especially the set about historical water temperature by hour for every single beach, one of his favorites. He cited this as a great example of microdata, with changes and patterns being "data that happens right in front of us."

These portals make data available to people so long as they have some experience working with the portal's interface, making it easier to search for data, filter, and download what's needed. Most of the work transforming the data into user-friendly formats has already been done for you. For more advanced users, the portals provide API keys from Socrata.

Searching and Scraping

The Searching & Scraping session of Data Days covered modes of getting data when there is no partnership or the data is not readily available. Featured speakers from Chicago's data ecosystem—Scott Robbin of Robbin & Co., Fernando Diaz, formerly of Hoy, Forest Gregg of DataMade, and Maryam Judar of Citizen Advocacy Center—discussed web searches, Freedom of Information Act (FOIA), and scraping methods to extract data. Below is a condensed summary of what they talked about.

> *"80% [of the work] is knowing what already exists."*
> — FERNANDO DIAZ, FORMER MANAGING EDITOR AT HOY IN CHICAGO

Boolean Operations

Boolean operations are powerful when applied to Google searches or when they're used in queries inside other search engines. The conjunctive logical operator "AND" returns values shared by two (or more) sources. The disjunctive logical operator "OR" returns all values from all sources, while the "NOT" operator removes values from a particular source.

When you're using a search engine, make sure to use an advanced search feature, if available, and look for indicators that represent Boolean operations. Some search engines might use =!, = =, -, <>, ˜, or NOT to represent "A NOT B".

Wildcards

In addition to "AND", "OR", and "NOT", many advanced search engines use wildcard symbols. A wildcard symbol allows you to specify a part of a word while leaving the end of that word up for grabs, meaning that if you searched "Redevelop*", the search engine would return records that contain the words "Redeveloped", "Redevelopment", Redeveloping", and so on. Again, be careful, since some search engines require different symbols and have different standards for wildcard searching.

Some search engines often use dedicated shorthand to describe records in their catalogs. For example, if you wanted to search just authors in the Internet Archive, you could use "AU ='Washington'" in your search. Common shorthand includes AU = Author, TI = Title, SO = Source, DE = Description. Bibliographic records contain all kinds of useful information, known as metadata, such as creator, origin, date of creation, media format, and so on.

Googling

Online searching can be a lot of work, but at the center of it lays a basic back-and-forth process: you make a query, expand the query results, and then refine the query for a new search based on what you learned from the first result list. You can limit your results by adding search terms, and then grow your results by following metadata hierarchies up into broader categories. Most of the time, you won't have a good idea of what your dataset will look like or where it will come from until you've found it.

Boolean logic, wildcards, and dedicated placeholders for common attributes (like AU for author or TI for title) can be used to refine your Google searches. The Google search engine can be used the same way as a library's advanced search engine.

Example

Let's say we're interested in Chicago Tribune articles written about a wave of Chicago Public School closures in 2013. If I Google "Chicago Tribune CPS closures" I get 51,000 results. But if I Google [site:chicagotribune.com "Chicago Public Schools" AND "Closures" 2012..2013] we get 163 results, all of which are from the Chicago Tribune's website and all of which relate to the recent school closures. The "site" operator allows you to specify which site you want to search, values in quotation marks will be your target text, and the ".." operator specifies a date range for your search. Explore Google's search operators to strengthen your searches and get access to data you want.

Scraping Web Data

But what if you already know where your data is?

Depending on the user agreement associated with an online data set, you might be able to scrape the data directly from an online source. Web scraping takes advantage of a markup language's underlying structure. Scraping is only as effective as how the structure indexes the website's data. By querying the website programmatically, you can extract the data most important to you.

Each entry listed in a table on a website, for example, has a corresponding HTML tag that distinguishes the entry as one element among many on the webpage. If you find the category that describes the elements in a table, you can use the name of the category in a program to generate a list of every item under the category.

Web scraping—and the work it takes to create a scraping program—might seem tedious to get at a table with only a few entries. Scraping becomes really valuable when you're working with tables that have thousands of entries, or if you need to query a large database that supports a website. Many object-oriented program languages, such as Python and R, have web scraping libraries.

Accessing data can be difficult. You have to know where the data lives, whether there are restrictions on using the data, and whether you can extract the data programmatically. All together, though, these skills make it far easier to access data you need.

References:

Forest Gregg has a great video tutorial on scraping with the Python programming language https://www.youtube.com/watch?v=yCcSP3GQhho

Gregg's tutorial also has a GitHub repository for reference https://github.com/fgregg/scraping-intro

A handy guide to 'Google-fu' https://en.wikipedia.org/wiki/Boolean_algebra#Diagrammatic_representations

Data Acquisition Session notes https://docs.google.com/document/d/1wwLUec1qTdb14VA538pd8Bkdy0OILNd-F_1CMANKXgg/edit?usp=sharing

Data Acquisition Session Video https://www.youtube.com/watch?v=kKxXNCrUoFE&feature=youtu.be

Regional Data Porals Session notes https://docs.google.com/document/d/1TVazX6JKYzI-yk5c4NqxmkSxCN-9LzIHXrDtI2FnMe4/edit

Regional Data Portals Session video https://www.youtube.com/watch?v=oxpOo7J4No4&feature=youtu.be

Searching & Scraping Session notes https://docs.google.com/document/d/1VdyyHkz5p3PKWKbumg7ZRP8JiVrmpqMeZxQuyocaGUU/edit

Searching & Scraping Session video https://www.youtube.com/watch?v=LT9Iyo88bVg&feature=youtu.be

On-Ramps

"It's about shifting the paradigm from consumer to creator."
—SANDEE KASTRUL, PRESIDENT AND CO-FOUNDER OF I.C.STARS

Many people want to benefit from and contribute to Chicagoland's data ecosystem, but don't have an opportunity to take that first step into the work. This chapter begins with a list of public meetups, where residents can learn skills and network. Then, this chapter will continue to discuss data ecosystem on-ramps for organizations and for young people—especially young people of color. Building on-ramps is some of the most challenging, yet crucial work to be done, since if the data ecosystem really works for people, it must include everyone's perspective, not just the perspective of a few. The ecosystem grows stronger the more people it can serve.

Meetups

Chicago has one of the most mature ecosystems focused on technology and skills building. Regular meetups, many through meetup.com, are key on-ramps into the data ecosystem.

Here's a list of Meetups that were talked about during the Chicago School of Data Days and some that have evolved since 2014.

- LISC Chicago Data Fridays
- Chi Hacknight
- DataPotluck
- Chicago City Data User Group
- NetSquared
- 501 Tech Club Chicago
- Chicago Counts!
- Hack At U Chicago

- Chicago Data Visualization Meetup
- R meetup
- The Data Scientist Chicago
- Blue1647 Meetup

Tech Training/Support Collaborations

> *"If you are not collaborating, you are leaving value on the table. This is the age of collaboration in the nonprofit sector."*
>
> —JEAN BUTZEN, THE PRESIDENT & FOUNDER OF MISSION STRATEGY CONSULTING, CHICAGO SCHOOL OF DATA DAYS

Many organizations continue to stress the lack of available resources for tech training and support within their current structure. A growing trend among organizations is collaborative sharing of expenses for back office operations. The Tech Training/Support Session at the Chicago School of Data Days, featuring Jean Butzen of Mission + Strategy Consulting, explored the strategic benefits of organizational tech-based collaborations and identify funding sources that support these types of efforts.

Example from Nashville

In 2010 the Nashville Chamber of Commerce released a Child & Youth Master Plan. They created a network made of 22 committees, a board of directors, 300 organizations, and 7 dedicated staff. They organized around a metric: High school graduation rate. The rate rose from 58% to 83% in two years. Truancy was reduced nearly 40%. These sharp changes in graduation and truancy rates were accomplished with a $1,000,000 budget. Note that many dedicated people contributed to the collaborative by volunteering their time and expertise. Many organizations contributed by folding the mission of the collaboration into their own work.

Organizations have to decide how the collaboration fits within their own missions, how it might affect their brand, how their employees are affected, and how the organization makes decisions on a day-to-day basis. Eventually, though, after all the work to make the collaboration concrete, it'll look like the collaboration between partners "just happened," meaning that the relationship between the organizations will become a regular part of all the staff's everyday work.

Given how straightforward collaboration sounds, it is a very challenging and complicated process. Many nonprofits have difficulty staying afloat, let alone being able to afford the investment in time and resources it takes to make collaboration work. Add privacy concerns between partners and the fact that lead organizations may change over time, and sometimes it seems like the challenges outweigh the potential value of collaboration.

Collaboration Models

During the session, Butzen described a spectrum of program integration. The further you got towards 100% integration, where basically one partner is taken over by another, risk increased. The middle zone, about 50% integration, was where the most opportunity and value could be found, and possibly the most reasonable amount of risk, too.

Butzen described four collaboration models that she believed to be most effective:

1. Intra-sector. A nonprofit/nonprofit partnership
2. Management Service Organizations. A group of organizations coming together, pooling the money they want to spend on services and jointly purchasing those services. This increases the quality of the management system and reduces cost. Since many nonprofits can't afford HR or IT services and staff members are doing 2-3 jobs, this model frees up staff members' time so that they can do what they do best. This model saves time and reduces expenses.

3. Shared Service Alliance. A hub and spoke model where the hub provides the administration as much as possible for the participants and others share services to a group of autonomous organizations. A Shared Service Alliance is also where organizations agree to share a particular service space, in part to share knowledge and reduce costs. For example, a foundation helped a group of Colorado daycares set up a central hub to facilitate training and marketing.
4. Cross-sector. A business/non-profit partnership

Butzen believed that the Shared Service Alliance model and the Management Service Organization model were especially valuable for members of the Chicago School of Data.

For the flow of money, there are three models:

1. unilateral flow, where a big company gives money to a small nonprofit
2. bilateral/parallel exchange, which both entities are equal in size and have an equal exchange;
3. conjoined resources, where each entity gives to each other, but is creating something new.

Although conjoined resources "is the most powerful collaboration," Butzen said that you want to have as many types of collaborations as you possibly can.

Choosing Partners

An audience member at this session asked, "How do you coach an organization?" Butzen suggested organizations start by answering these questions: What are you trying to accomplish? Where are you stuck? What is causing environmental barriers?

For example, if someone is interested in growing but doesn't have the resources, look at who is out there and who you would want to grow with. The book James Austin's Creating Value in

Nonprofit-Business Collaborations was recommended as a good resource for organizations who want to learn more.

In finding partners, Butzen recommended looking at your missions and objectives, values and motives, your strategies, and make sure they're clear to each partner. It's okay if they're not entirely the same. "What's different about the partner might be what's good about the partner," advised Butzen. Performing a strengths, weaknesses, opportunities, threats (SWOT) analysis of the partner is advised. If you've got multiple prospects for partnership, rank and evaluate them on these categories to help guide your decision.

More advice included:

- You should be looking for partners you trust, perhaps someone you've already worked with.
- Some part of your vision, mission, or strategy should or could be shared.
- Definitely make sure that you share the full scope of the partnership internally with your own organization.
- Any joint planning henceforth should be put in writing.

Diversifying Competitiveness in Technology

This session explored the timing, availability, and opportunities of technology on-ramps for youth in Chicago and what it will take to influence a paradigm shift by 2018. It featured leaders and workers in the midst of making this change: Laura Sanchez, Emilie Cambry, and Sandee Kastrul. Sanchez is the CEO of a company named SWATware which is based in the South Side of the city. SWATware seeks to be an "external IT department for local businesses" who are incapable of solving computer problems that arise for themselves. Cambry is the founder of the coworking space and incubator, Blue1647. Kastrul is the President and Co-founder of i.c. stars, a technology education center.

These leaders came together with conference participants to explore what technology on-ramps are available for Chicago youth.

Smart Chicago's own Kyla Williams moderated the panel. Four general strategies were discussed: amplifying youth voices, providing mentorship opportunities, empowering through entrepreneurship, and digital/data skill-building for future success.

Amplifying Youth Voices

Too often youth voice is left out of conversations among policy-makers and leaders in technology. Youth voice is an important way of increasing diversity in technology. Of course, bringing youth voice to the table in just a token way, without really engaging youth, does not do justice to the youth perspective.

One way of getting young people excited about technology is to start teaching technology earlier in school. Both Williams and Sanchez argued that tech training needs to start much earlier for young people. As Sanchez said: "We need to start with elementary or even early childhood education. In high school, the geek isn't cool. We need to change the perspective and mentality to get more diverse people into IT." How does the ecosystem make sure young people access the on-ramps built for them?

Providing Mentorship Opportunities

Mentoring relationships, especially near-peer mentoring, are extremely powerful in driving diversity in the technology sector. As Kastrul said: "The best mentors are the ones who can see us for who we are and who we can be." Relationships of reciprocity can last for decades. To create matches between mentors and mentees, i.c. stars, for example, used a model like the television show "The Voice," where mentors turns around in their seats and listen to a 2-minute presentation from potential mentees. Then they turn back around, and the mentor makes a match. The goal of these mentorships is to help young people and their mentors thrive in all of their pursuits.

Entrepreneurship

Kastrul reminded the participants: "Nothing stops a bullet like a job." Civic leaders and business leaders need to teach entrepreneurship and develop businesses in communities of color. When conversations happen across sectors, through collaboration, on-ramps emerge and silos break down.

Cambry discussed a partnership with 500 churches to link social enterprise with digital training. Organizations could pay youth $500 for a project that a developer might charge $1,500 for. Or, instead of paying for a staff member, a network of organizations could outsource their development work to a group of young people, similar to a Shared Service Alliance, with young people at its core.

Skill-Building for the Workforce

> *"Those of us who have overcome things—we have skills. We need to stop the narrative that we are needy when we are really warriors. We are experts at solving problems...Learning technology is the easy part."*
>
> — SANDEE KASTRUL, I.C. STARS

Increasing diversity in technology is crucial for the ecosystem's success. At the time of the conference, Blue1647 had just finished workforce development training for its first cohort of 90 young people. Their pilot program was immersive, and 90 young people learned HTML, CSS, JavaScript, and JQuery. They created GitHub accounts and developed their own digital portfolios. Projects included games, apps, and websites. Ideally, with these new skills, young people could build websites for small business and nonprofits in Chicagoland.

"We're trying to convince kids that spending 30 hours a week learning about technology is a worthwhile investment," Cambry said. Sanchez agreed, saying, "We need to create long term goals for community growth."

References:

Meetup Session notes https://docs.google.com/document/d/1AoN-B_1H5pTRSuqlZnzLymMVC2R-E9dVL-7iDjREDhg/edit

Tech Support / Collaborations Session notes https://docs.google.com/document/d/1q-uvQv7u68UujlDO_yzt9fPt6r-hsoOpZO9Msm-vjuw/edit

Diversifying Competiteveness Session notes https://docs.google.com/document/d/1nJLZu3EhbfgsoJvokuqWd8WY3fnBT-_CbNxDSkcpMQI/edit

Diversifying Competitiveness Session video https://www.youtube.com/watch?v=g5KFezWil7k&list=PLJ75D_m2b5GtN9bb5ZT6y4ggI8dR4TtX-j&index=18

Tools

"What does the community need? What does the community want? We will never decide in this room, between you and I, what we're going to do as an organization. We let the community tell us what it needs and then we respond to it. Yet, I think we still need data for that."

– JAMES RUDYK, NORTHWEST SIDE HOUSING CENTER, CHICAGO SCHOOL OF DATA INTERVIEW WITH MATT GEE

There are many tools available to support all parts of the data pipeline—tools to collect, manage, analyze, and publish data. Many tools in the ecosystem are free and open source, so that you can access a tool's source code and get full control of its functions.

According to the Chicago School of Data Survey, these were the tools most used by the ecosystem:

- Desktop spreadsheets (231 organizations)
- Online spreadsheets (164)
- Website data analysis (138)
- Online surveys (179)
- Proprietary customer relationship management (CRM)/database tools (132)
- Open source CRM tools (17)
- Open source databases (55)
- Open source data analysis (40)
- Proprietary analysis programs (52)
- Proprietary data visualization tools (40)
- GIS and mapping tools (79)

Based on the survey results and supplementary interviews, we found that the top data tools used by organizations were spreadsheets (both on desktop and online), web-based data analysis tools, online surveys, and proprietary CRM/database tools. We also identified three sessions within the broader "Tools" category that would interest the conference participants: Cleaning Data, Collecting Data, and Mapping Data.

Cleaning Data

The Cleaning Data session focused on tools and methods to clean data collected and maintained in the desktop and online spreadsheets—the most popular tools in the ecosystem. Sometimes the hardest part of working with data is error correction. Cleaning data is an important step in getting data to work for you. David Eads and Geoff Hing led the session.

Hing likened the data cleaning process to being a janitor. He gave a broad-level overview of the data cleaning pipeline. Eads described the data cleaning process through a case study about NPR's article "MRAPs And Bayonets: What We Know About The Pentagon's 1033 Program."

Working with criminal records in Cook County, Hing found misspellings and different encoding systems that needed metadata description. He often has to combine two values into a single column with concatenation functions.

Common problems with "dirty" data

- Misspellings
- Combine two values into a single column (concatenation)
- Coding systems discrepancy due to changes in codes over time
- Encoded values without metadata explanations

Geoff Hing reminded the audience, "Understand data before you start cleaning." Sometimes there are encoded values that have a special meaning that you may not be aware of. One example he gave was an eight-digit column that had values like '5, 90, 24000, 10, 30000, 14,' and it really was signifying time. For this reason, it is great to have a data dictionary.

Several important cleaning tips shared by Hing and Eads

- You should know *how* the dataset was created. Understand the workflow; test the data acquisition process from beginning to end for "friction points" that might generate messy data.
- Do a visual inspection of the spreadsheet, look for empty columns, and scan for any values that stand out as strange. Sort the columns to help identify those outliers.
- Be sure to keep all original data values. Don't edit the original values.
- There are various toolkits available to clean your data like csvkit, custom scripts in Python, and OpenRefine. Or you can clean data directly in the spreadsheet.
- Document the cleaning you've done and then replay the process to verify its effectiveness.

Creating a Data Pipieline

Hing and Eads emphasized the importance of creating a data pipeline. With a pipeline, you can automate the data cleaning process with a scripting language, which in turn makes it easier to manage versions of your dataset from importing, summarizing, and exporting. This is most clear when you use version control, such as through GitHub, to keep track of the workflow. Along with csvkit, OpenRefine, and Python, Eads also uses Pentaho, Excel macros, and Anaconda for data cleaning.

Collecting Data

This session covered different modes of collecting and storing data in various systems. Dr. Lance Kennedy-Phillips, formerly of the University of Illinois-Chicago, Anne Cole from Neighborhood Housing Services of Chicago, and Smart Chicago's former Executive Director Dan O'Neil led the conversation. They highlighted ways that their organizations approached and thought about data collection.

Kennedy-Phillips focused on the broader field of institutional research and wanted the audience to know about valuable secondary sources for data about higher education. He divided the datasets into local, statewide, and federal. He mentioned several other datasets, listed under resources, but emphasized that the data in UIC's enterprise system is designed around custodians, who collect data about students, producers, who create the reports, and the users, who make the policy decisions.

Cole discussed the challenges of collecting data from the ground up for nonprofits. The Neighborhood Housing Services of Chicago, which served 6,000 people in 2013, is trying to build a data warehouse for their client-side data and their loan-level data. Surveys are an important interface between the organization and their clients, with the goal of keeping track of their clients over time. Their data ultimately gets used for reporting and public policy outreach. Quarterly, the organization meets internally to discuss how well their data strategy is working. Ultimately, they want to streamline their data collection process to support their administration and to bolster their funding.

Cole described the steps her organization took to create the data warehouse. First, they inventoried and aligned all their data sources from the different organizational levels, which were siloed in Excel spreadsheets, rogue Access databases, and in people's brains. The end goal of this first step was the creation of a data dictionary. Second, they developed the data framework with their regular legal reporting in mind, so that they could automate the creation of these reports. Third, Cole described how her organization had to learn how to overcome capacity limits in order to get their warehouse off the ground.

Mapping Data

Maps can literally "ground" data, presenting it in a functional and accessible way. During the "Mapping Data" session at the School of Data conference, we learned about some simple tools to create maps quickly—Google Fusion Tables, Searchable Map Template, QGIS, and more. Derek Eder of DataMade, Mike Reilley of the Red Line Project, and Josh Kalov, Smart Chicago Consultant, led this session.

Building on Open Government Data

Over 600 unique datasets are free to view and download in a variety of formats on the City of Chicago Open Data Portal. Cook County maintains a similar site. Datasets can be exported in .kml formats and uploaded into a Fusion Table. Derek Eder is an open web developer, owner of DataMade, and ChiHack Night leader, created a searchable map template using Google Fusion Tables. Eder provided a demo and instructions on his website, derekeder.com.

Eder also showed us an example he created with Open City: the Vacant and Abandoned Building Finder. This site maps empty buildings across Chicago, with optional filters to see neighborhood demographics relating to poverty and unemployment rates, income, and population. The site also provides information on reporting abandoned buildings.

Telling Stories with Maps

Mike Reilley is the founder of the Journalist's Toolbox. As a professor at DePaul University, he also founded and advises the Red Line Project, a news site that covers Chicago neighborhoods located near CTA red line stops. Reilley used mapping software to create a map of homicides in the city, as well as the Chicago Stumbler, documenting poor sidewalk conditions. Reilley had the idea after he turned his ankle on a cracked sidewalk. Using this data, he was also able to post calls to action, enabling residents to tweet and email their aldermen to complain.

Reilley also showed a StoryMap that the Redline Project had created for the 2014 FIFA World Cup. StoryMap is a free, open source tool created by Northwestern University's Knight Lab. You can annotate maps with photographs and other media, information, and links. This is a great tool for journalists and storytellers who are interested in presenting information that is embedded in particular geographies. StoryMap has been used by the Washington Post, Al Jazeera, CNBC, Foreign Policy, and other publications.

Open Source Mapping Tools

Josh Kalov shared alternatives to ArcGIS, the professional desktop software that is often purchased and used to create maps. QGIS is a free and open source alternative to ArcGIS, and unlike the online software Eder and Reilley mentioned, QGIS can create printable maps in PDF and JPEG formats complete with legends and annotations. QGIS runs on Linux, Unix, Mac OSX, Windows and Android systems. Though QGIS is free, Kalov pointed out that its user interface isn't as friendly as ArcGIS, and its reliance on plugins can be a drawback.

Audience members brought up other tools for map makers, such as Mapbox, a free and open source software for creating custom maps, and Power Map for Excel, which is available for 2013 MS Office users as an unsupported add-in, or free for subscribers of MS Office 365.

Note that you can access a full list of categorized tools and resources from the Chicago School of Data in our Resources Chapter.

Resources

David Eads's presentation slides "Data Processing Workflows"

Cleaning Data session notes https://docs.google.com/document/d/1RlK-gNzsoZHdPudAdNX16aMG5NFZKdKRYm4V6Axquj-4/edit?usp=sharing

Cleaning Data session video https://www.youtube.com/watch?v=K-tayrGSgE1Y&feature=youtu.be

Collecting Data session notes https://docs.google.com/document/d/1ti-Ukka9Gaa3WBkrW8zjlAVM6rjpTn_pY60H7kzR00DQ/edit?usp=sharing

Collecting Data session video https://www.youtube.com/watch?v=GHmuzo-OLNg&feature=youtu.be

Mapping Data session notes https://docs.google.com/document/d/1M7U03HeodoVuJiLfRlgaONcjT3vBDYAzUBD8asp5MN0/edit?usp=sharing

Mapping Data session video https://www.youtube.com/watch?v=GHmuzo-OLNg&feature=youtu.be

QGIS tutorial http://maptimeboston.github.io/qgis-101/#0

Derek Eder of DataMade showcasing the Vacant and Abandoned Building Finder at the Chicago School of Data Days *(Photo by Stephen Rynkiewicz, Chicago School of Data Documenter)*

Current State of the Ecosystem

The Chicago School of Data aimed to convene and shine light on Chicagoland's vibrant data ecosystem—the thriving people and organizations utilizing data to advance the public good every day. In the last several chapters, we've overviewed the insights and best practices shared by various data practitioners at the Chicago School of Data Days.

Before those leaders were convened, research was conducted at a higher level to understand our data ecosystem's basic characteristics, strengths and challenges. That scan of the field was informed by interviews as well as our census form and a summary of our findings is captured in this chapter.

A Taxonomy of the Ecosystem

Creating a framework to center our conversations about this work and this field was an essential prerequisite to the Chicago School of Data. This framework, or taxonomy, equipped us with a structure and language as we took inventory of and convened the community of practice.

Smart Chicago tasked consultant Andrew Seeder with creating a working taxonomy for Chicago's regional data ecosystem. The trial and error involved in his process clarified the project's perspective. For example, in Andrew's words:

> *Originally, we classified Chicago School of Data participants by industry. Participants were either a university, government department, nonprofit, or private company. These buckets were useful when we were finding people to interview. We wanted to get as broad a*

cross-section of the landscape as possible. We didn't want to miss the perspective of any of our partners. Over time, though, we found that these buckets weren't specific enough for our purposes. They weren't organized in a way that told us anything about how our partners really used data. Ideally, we wanted look at an organization's place in the ecosystem, its niche, and know exactly what support it needed around data and how else the organization could benefit—and be benefited by—the ecosystem.

Throughout the project—before the survey, after the survey, and after Data Days—Andrew revised this taxonomy to help us to characterize and categorize the players in this space. The full structure of the data ecosystem is public and posted on GitHub for others to use and improve. It has gone through several revisions, most notably making the categories within "data consumers" more specific.

The Chicago Data Ecosystem has:

- Creators
  - > Who open their data for free
  - > Who open their data for a price
  - > Who don't open their data
    - Because of technical capacity
    - Because of cost
    - Because of legal agreements
    - Because of the public interest
    - Because of other reasons
- Consumers
  - > Who only consume free data
  - > Who pay for some of their data
  - > Who use data for
    - Resource allocation
    - Measuring impact

- Advocacy and outreach
- Understanding the needs of people served
- Donor development
- Operations
- Research

• Enablers who provide services and goods in the ecosystem such as

> Volunteers

> Consultants

> Funded organizations

> Paid organizations

You can see more pieces of the taxonomy: https://gist.github.com/danxoneil.

We anticipate that this framework will be improved and refined more over time as we learn more — and as the ecosystem evolves. Regardless, this is a point to understand the challenges and opportunities for growth in Chicago. We also anticipate that this framework could be of use to other cities seeking to map out their own data ecosystems.

Characteristics of the Ecosystem

The diverse data ecosystem in Chicago includes multinational corporations, tiny local nonprofits with shoestring budgets, philanthropic institutions, small companies, and public agencies.

Our region's data ecosystem is supported largely by data created (and released) by a few large public enterprises. Among data creators, some organizations provide their data for free, at no charge to either the public or other organizations. These organizations include the City of Chicago, Cook County, and the U.S. Census Bureau — large institutions that fuel the work of researchers, developers, and advocates in the ecosystem.

How are members of the ecosystem using data?

- Resource allocation (133)
- Measuring impact (210)
- Advocacy and outreach (179)
- Understanding the needs of the people served (199)
- Donor development (125)
- Operations (138)
- Research (175)

Some organizations create packaged data sets of data they've collected, while other organizations make it a business of cleaning free, public data. Others donate hardware and their expertise to local schools or, as an institution, they fund organizations working in the field. But data creators consume data and data consumers enable others to create data. These broad categories aren't mutually exclusive.

Challenges for the Ecosystem

Chicago benefits from the size of its data ecosystem and the diversity of players and experts. There are also several platforms on which residents, researchers, institutions can access and interact with neighborhood data: Plenar.io, OpenGrid.io, Chicago's Open Data Portal, as well as the websites of key data and research organizations like the Woodstock Institute and the Heartland Alliance. Also, while some cities might have only a handful of organizations who collect data, analyze data, build on open data, and train in the use of data, Chicago sees many organizations and events dedicated to these functions —some of which are explored earlier in this book.

There are natural challenges that accompany being a large vibrant, data ecosystem. It can be difficult to coordinate and communicate. For instance, there are ecosystem-wide functions that could add value for everyone: cataloging available data, convening players, facilitating partnerships, and incentivizing research across neighborhoods and policy specializations.

In the National Neighborhood Indicators Partnership's assessment of Chicago in 2013, they recommended a model to help coordinate the ecosystem:

> *In addition to the obvious data and technical capacities, the organization that houses the new activities needs to be considered neutral, have the respect of the broader community, and be oriented to practitioners. The office would need to have a holistic perspective—across topic areas, geographic levels, and municipal boundaries. To increase the chances of success, the work of the office should reinforce or enhance the other goals of the organizational home. The office staff need to be familiar with availability of data resources across topic areas but, in addition, must have the ability to work in a respectful way to help non-technical audiences access and use community indicators.*

Challenges for Organizations Within the Ecosystem

In terms of organization-level challenges, The Chicago School of Data Survey results pointed to several common difficulties. When asked about challenges with using data, the 259 respondents chose 1-8 characteristics that applied to their work:

1. Unable to dedicate the time to data work given other demands (141 respondents selected)
2. Staff lack the necessary technical skills (computer literacy, data management, data analysis) (110 respondents selected)
3. Unable to gain access to the data we need (79 respondents selected)
4. Unable to afford the tools we need to make use of data (69 respondents selected)
5. Not sure where to find or how to collect the data we need (68 respondents selected)

6. There is no software/website/tool that meets our data needs (48 respondents selected)
7. Other (36 respondents selected)
8. Not sure how data can further our mission (19 respondents selected)

Fifty-one respondents to the Chicago School of Data survey didn't select any of the challenges above. Predictably, respondents in this category were more likely to have used proprietary data analysis software and all knew how data fit into their missions. By comparison, 62 respondents of the 259 marked 4 or more of the listed challenges. Over 50% of this group self-identified as lacking "basic data literacy skills" and 9 of the 62 in this group lacked computer skills. The group experiencing more of the challenges was more likely to say that it was unsure how data would further its mission.

This gap in data access, skills, and tools was not only noticeable in the survey results, but also from the Chicago School of Data Days sessions. There were many organizations with the desire to contribute to the ecosystem, but they lacked internal technical capacity, funding, time, or they needed pre-existing legal agreements to grant them access to needed data.

For more information, see the public raw responses from the 2014 Chicago School of Data census.

References

http://www.slideshare.net/smartchicago/chicago-school-of-data-census-formoctober282014

http://www.smartchicagocollaborative.org/key-takeaways-and-raw-responses-to-chicago-school-of-data-census-form/

http://www.urban.org/sites/default/files/alfresco/publication-pdfs/412784-An-Assessment-of-the-Community-Information-Infrastructure-in-the-Chicago-Metropolitan-Area.PDF http://www.urban.org/research/publication/case-study-open-data-chicago/view/full_report

Raw census responses https://docs.google.com/spreadsheets/d/1ALP5vZCwkf6hNn8BH_UNY3IwDxHTeCAm7JAWVTPyy20/edit#gid=0

Conclusion

It's been three years since the Chicago School of Data Days conference, and in that time, the field has continued to evolve and expand. To know how to serve and support the civic data practitioners of Chicago, we hope to periodically measure and monitor the ecosystem, its players, and its needs. More interviews, convenings, and surveys should be conducted. It's clear that more partnerships can be facilitated in our community—especially those linking community-based organizations to Chicago's data intermediaries and data sources.

According to the Urban Institute: "Local data intermediaries are data translators, educators, conveners, collaborators, and voices for change. They use data to describe their communities, and they empower communities to use data in their activities, from community building, to advocacy and program planning, to policymaking."

The Chicago School of Data Census and the Chicago School of Data Days in 2014 were documented in this book for the health of our ecosystem, but the work has only just begun. While the Chicago School of Data census and the Chicago School of Data Days Conference convened an important community and spurred important conversations, we need to take next steps to sustain those conversations and turn those conversations into action.

To conclude this book, we wanted to build on the wisdom collected from partners during this project with an inventory of recommended next steps. Below are a few ideas we can take on to strengthen our data ecosystem.

Data Training

The Chicago School of Data census and Data Days conference revealed the need for more data training across organizations. Fortunately, there is a plethora of existing data training services in

Chicago. From the ChiHackNight, LISC Chicago Data Fridays, to Blue1647, many respected organizations hold regular events and training, both paid and open to the public. Also, during the conversations about "On-Ramps" to our data ecosystem, we learned about Chicago's vibrant Meetup community.

In a decentralized data training system where many are doing excellent work but none are responsible to meet the needs of all, how are we doing? One important task, and a possible future next step for the Chicago School of Data, will be to answer that question—by thoroughly scanning training offerings across the city.

Here are some questions we can ask ourselves about the holistic state of data training throughout Chicago:

- Payment: Is the training free or paid?
- Level of Service: Targeted to building individual, neighborhood, or organizational data capacity?
- Focal Point: Are the trainings mostly tool-centric or problem-centric?
- Timeline: Are trainings mostly one-time, standalone events or sustained?
- Training Format: Are trainings casual and drop-in or more formal?
- Training Motivation: Does the training start with open data or does it start with a mission/challenge?
- Geography: How spread out is data training? How accessible is it?
- The Training Ladder: Are all skill levels being served? Is there a visible pathway of training services in Chicago?
- Supply vs. Demand: Do the current collection of trainings in the ecosystem address the discovered challenges, preferences, and needs?
- Awareness: How known are the current training opportunities?

In the "Current State of the Ecosystem" chapter, we shared the eight challenges listed on the Chicago School of data survey. While more data sharing, open source tools, and open data in general can address many of the ecosystem's challenges, the most prevalent identified challenges —"Unable to dedicate the time to data work given other demands" and "Staff lack the necessary technical skills" can only be fixed with investments in people. How can the ecosystem efficiently train up mission-driven people within the resource-strapped organizations that make our communities and cities stronger? Asking and answering the questions above is a good starting point.

More Support Services

Some of the most informative and impactful data organizations in the city are specialized researchers like World Business Chicago, Chapin Hall, Heartland Alliance, Woodstock, the DePaul Institute of Housing Studies—just to name a few. Their work has positive ripple effects across the ecosystem. Mission-driven and data-driven, these organizations provide an essential function in the data ecosystem that ensures that information can become actionable.

Here are questions worth asking: How can this layer of our ecosystem be better supported? How can services be provided or pooled to help them?

Potential ways to support the work of Chicago's anchor data organizations

- Cataloging and inventorying available data across sources for practitioners
- Project-based funding incentivizing collaborative, cross-disciplinary work
- Usability testing services of new data tools, websites, and products
- Encouraging new collaboration models (see "On Ramps" chapter for more ideas)

- Civic engagement programing to link communities to new research or tools

This is a list worth thinking about more and potentially expanding. This idea of centralized or ecosystem-wide services is something we can build together.

More Communication & Collaboration

After the Chicago School of Data Days we asked participants, "What would you like to see as an outcome of the Chicago School of Data Days?" Here are some answers:

- "Greater coordination across efforts"
- "Increased communication between organizations that cultivate streamlined processes and a more cohesive network of supported colleagues."

More communication and coordination across organizations is low-hanging fruit. It could facilitate more collaboration (possibly untapped or unexpected partnerships), amplify great work happening across the city, and reduce any redundancies in our ecosystem. Most importantly, it would give support and community to the workers in this field.

Methods for increasing community-building across the ecosystem

- Having a source/place/platform for regular communication, convening, storytelling and data sharing.
- Awards/recognition for excellent or innovative use of data to build a better Chicago with residents
- A regular census of the ecosystem to measure changes in gaps and challenges
- More convenings

- Common onboarding into the community of practice—an easy, accessible step for mission-driven organizations using data to link up to conversations, network with partners, and link to training opportunities across the city.

We look forward to taking next steps with and for all partners interested in continuing this journey. Please contact us at info@smartchicagocollaborative.org.

References

https://docs.google.com/spreadsheets/d/1ALP5vZCwkf6hNn8BH_UNY3IwDxHTeCAm7JAWVTPyy20/edit#gid=0

http://www.urban.org/research/publication/nnips-guide-starting-local-data-intermediary/view/full_report

Alaina Harkness, then Program Officer of the John D. and Catherine T. MacArthur Foundation, addresses Chicago School of Data Days participants at the September 19, 2014 reception at the JW Marriott Burnham Ballroom *(Photo by Daniel X O'Neil)*

Meta: how we did it

Since 2014, we've been pleased to see information and resources from the Chicago School of Data appear in Data Smart Cities, the National Neighborhood Indicators Partnership website, and the City of Chicago's Tech Plan. We've also been pleased to see other organizations like BetaNYC adopt similar ecosystems projects.

Given the questions and interest we receive about replicated pieces of this project, we wanted to record and share the decisions, models, processes, and philosophies that fueled the Chicago School of Data. This is the "how-to" of our efforts to map, describe, align, and build capacity among the organizations and individuals working to use data to inform urban development in Chicago. This chapter is part cookbook, part history. We hope other organizations can use what we learned to organize their own Data Day.

Project Components

There were three main components to the Chicago School of Data:

1. Convene a core group of practitioners in Chicago who are using data to improve the lives of regular residents
2. Document and map the landscape of data activity in Chicago—the entities, tasks, companies, enterprises, civil service organizations, and others who make up the field
3. Plan a region-wide event where we share this mapping work with the larger data community and showcase all of the activity underway through capacity-building workshops and demonstrations

The project began by hiring a research consultant who helped us plan our interview agenda. The interviews focused on practitioners in the field, although the interviewees represented diverse perspectives. The interviews guided the design of our conference agenda. And, along with our interviews and conference outreach, Smart Chicago sent hundreds of organizations a "data census," which further refined our conference agenda. The project culminated with a two-day conference, Chicago Data Days, held in Willis Tower on September 19 and 20, 2014. You can see lots of documentation about our conference in recorded sessions, shared meeting notes, and images.

Project Philosophies

> *"We literally can't get anything done without other people."*
> —DANIEL X. O'NEIL, FORMER EXECUTIVE DIRECTOR OF THE SMART CHICAGO COLLABORATIVE

The warrant for Smart Chicago's "ecosystem project" is to build with, not for, to be at the service of people. We wanted to learn from people all over Chicago working with data and data tools to further their public missions. The intention was to craft a project and conference that was shaped by what they told us. Smart Chicago did not come with a framework in mind for this project or the conference. It was built with the people we spoke with. We want to capture that in the how-to of what we did.

Individuals and organization could engage with us through several modes:

1. an open form that anyone could fill out
2. an in-depth interview when requested

Inclusion was another key philosophy. Anyone who wanted to participate in the School of Data could and was encouraged. Smart Chicago was not prescriptive about what "types" of people or organization we thought could contribute to or benefit from this work. The result was a process that was open, welcoming, and left room to find unexpected value.

Funds

The Chicago School of Data was supported by a $250,000 grant from the MacArthur Foundation which was given to "support [Smart Chicago's] Chicago School of Data project, an effort to map, describe, align, and build capacity among the organizations and individuals working to use data to inform urban development in Chicago."

Timeline

From early December 2013 on, the project has had several installments, beginning with over 40 in-depth interviews, organizing a conference around a "Data Census," and ending with the publication of this book. Oprima-1 conducted outbound research to gather as many practitioners as possible. On September 19th and 20th, these themes and talks culminated in the first Chicago School of Data Days.

Our Tools

Here is an overview of the tools we used to undertake this project:

Eventbrite: We used this leading ticketing tool for managing conference attendance. We had an event page, and we got individual emails every time someone signed up to come to the conference on September 19 and 20, 2014. We used the Eventbrite API to pick up all of the information from event signups and wrote that to our master list.

Mailchimp: We used Mailchimp to manage our communication to potential and actual attendees. It has a great API, great analytics, and easy-to-use email templates. Their segmenting tools are great, so we were able to send different emails to people who were already signed up for the conference and people who had yet to get started with our initial data census form.

Wufoo: Wufoo was one of the most essential outside tools for us during the Chicago School of Data project. It fueled the Chicago School of Data Census as well as the form that recruited Smart Chicago Documenters. It helped us gather so much information, and it has a great API that allowed us to pull that information into other systems we maintain.

WordPress: We use WordPress for many websites, including Smart Chicago's. We made the Chicago School of Data Days conference website using WordPress, with plugins and APIs to use data we collected from other tools.

Dropbox: Smart Chicago has a Dropbox for Business account where we keep all of our key documents and do lots of sharing with our core stakeholder team. The documentation from our Smart Chicago Documenters was also centralized on Dropbox. We also used Dropbox integration with Wufoo to allow organizations to upload examples of datasets and reports that they use in their daily work.

Google Docs: We use Google Docs and Sheets mainly for project management and joint composition of things like the agenda and draft emails.

Evernote: We used this early on for notetaking during some initial interviews.

Microsoft Excel: We are working with Oprima-1, a bilingual call center, to do research on Chicago organizations that use data to make lives better. They sent us daily reports on their work in Excel.

The Chicago School of Data Census

The census form was a key part of shaping the Chicago School of Data project and the conference. We took responses from this survey to better understand how data was used by organizations and responded by creating themes that defined the conference sessions and discussions.

The form was shared often and was open publically — you did not need an invitation to fill one out. We got 259 responses. Here they are, with identifying information and end-matter (re: how they want to be contacted, participation in the project, etc.) removed.

In terms of tools, we used Wufoo to create the form and a combination of phone calls, blogging, in-person outreach, and email campaigns via MailChimp to spread the word. We used Zapier to connect the Wufoo form to Google sheets so that results could be shared publicly.

Interviews

As we distributed the Chicago School of Data census, we followed up with in-depth interviews of organizations—a deeper dive that helped us discover and discuss opportunities, and understand where we could bridge gaps. To do the interviews, we partnered with Adam Garcia of Nectel Technologies.

During these interviews, we asked organizational representatives what they did, who they served, what data they collected, how they use data, and their capacity to collect and use data.

A list of the interview questions we used

- What does your organization do?
- What do you do?
- How does data drive what your org does?
- Can you give an example of one of the most successful examples of this?
- What communities does the org serve?
- How do you discover the needs of communities you serve?
- How do you measure your impact?

- What have been your organization's biggest challenges in working with Chicago's data ecosystem?
- How do you measuring the value of the services you offer or the information you provide?
- What are your organization's most important sources of information?

Outreach

E-mail Campaigns

As mentioned before, we invited everyone—especially at the beginning of this project. We did not want to make assumptions about who was in the data ecosystem or not, regardless of their work, experience, or challenges. Our first email campaigns, implemented and organized through MailChimp, emphasized the philosophy of inclusion.

Here is a history of our email campaigns for the Chicago School of Data, from introduction of the project through conference follow-up:

1. "Chicago School of Data Outreach" to 177 subscribers
2. "Chicago School of Data -- Terry Letter" to 345 subscribers
3. "Chicago School of Data Register Email to Form Fills" to 82 subscribers (who filled out our initial form)
4. "Chicago School of Data -- No Initial E-mail Sent" to 628 subscribers (who did not receive an initial email from Oprima-1 outreach)
5. "Chicago School Of Data -- Expand To Networks" to 158 subscribers
6. "Chicago School Of Data -- Meetup Row" to 25 subscribers who are part of meetup groups
7. "Conference Details For Attendees" to 275 subscribers
8. "Last Minute Conference - Attendees" to 346 subscribers
9. "Data Days Conference Follow-up" to 346 subscribers

Oprima-1

We hired Oprima-1 to help us conduct research and data gathering for our Chicago School of Data Survey. They conducted 50 hours of calls and follow-up calls with organizations across Chicago that use data.

Total count on the following fields from Oprima-1's research

- Individual Twitter Handles: 199
- Organization Facebook: 1574
- Organization Twitter: 1158
- Organization URL: 1754
- Organization About Page: 1711
- Contact Emails: 1078

In the course of their work, Oprima-1 often made the first point of contact with organizations, introduced the Chicago School of Data to them, and then followed up with an email with more information.

Partner highlight: Oprima-1

Oprima-1 is a program aimed at people who want to learn English and also become professionally trained to work in call center customer service. Executive Director Daniel P. Loftus is a great leader in the field of digital skills, and Operations Manager Gricelda Piedra leads the team that always provides us fast, excellent service.

The Chicago School of Data Days Conference

Participants

See the list of participants in the Participating Organizations chapter.

Speakers

An essential ingredient to the Chicago School of Data was attracting exciting, and compelling speakers to participate during Data Days.

Here is an example of a note that we used to initially recruit speakers:

"Dear [FIRST NAME],

I am writing to invite you to speak at the Chicago School of Data Days conference held on September 19 and 20, 2014.

Here's info about the conference session we'd like you to be a part of:

Session Name, Day, Month Date from starting time - ending time

Session Description

This session is one of three morning/afternoon sessions designed to foster discussion about the greatest challenges organizations currently face in using data.

In your survey, we learned about <your organization>. and how you use <type of data>. While the City is a leader in publishing data, there are always datasets that remain unpublished, and we would love if you can talk more about your experience trying to access data.

We expect to have three speakers total, all of whom we are in the process of inviting. Each of the three speakers will present for about 15 minutes.

Then you and your fellow presenters will help facilitate discussion about the topic. Based on the feedback we've gotten in our work to date, we expect others from the audience to share their experiences as well. This is just one way we hope to move away from a traditional Talking Heads panel and allow for interesting and invaluable dialogue around this topic.

We're also publishing lots of the raw insights and analysis via blog posts, audio snippets, and other methods before the conference. Most importantly, this session and the entire conference itself is going to deeply affect our final report— a scan of the field that will serve as a model for data ecosystems across the nation.

Please let me know if you are available and interested in serving as a speaker and can make it at this timeslot. I really hope you can make it!"

We found our speakers in a very interesting way. Instead of finding the "usual suspects" when it comes to speakers at data conferences we were looking for the "workers" in the field. Therefore, we looked at the surveys and interviews and invited people who shared with us to then be part of the conference as speakers.

List of Chicago Data Days Speakers along with their titles/institutions at the time of the conference:

- Brenna Berman – CIO, City of Chicago
- Kathy Pettit, NNIP
- Jill Young, Directory of Research and Evaluation, After School Matters
- Margaux Pagan, DonorFuse
- Taryn Roch, LISC Chicago
- Samia Malik, Chatham Business Association
- Spencer Cowan, Woodstock Institute
- Stephen Pigozzi, Association House
- Eliza Moeller, University of Chicago
- Traci Stanley, Christopher House
- Tracy Siska, Chicago Justice Project
- Andre Kellum, 741 Collaborative
- Kathryn Bocanegra, Enlace Chicago
- Nate Steinfeld, ISPAC
- Matthew Roberts, Chicago Department of Public Health
- Vivian Hessel, LAF
- Matt Bruce, Chicago Workforce Funders Alliance

- David Eads, NPR
- Dan Sinker, Knight-Mozilla
- Dan O'Neil, Smart Chicago
- Beckie Stocchetti, Kartemquin Films
- Chris Hagan, NPR
- Emily Withrow
- Joe Germuska
- Sarah Duda, IHS
- Susan Yanun, Logan Square Neighborhood Association
- Sean Vinck
- Simona Rollinson
- Derrick Thomas
- Tom Schenk
- Scott Robbin, Robbin & Co
- Fernando Diaz, Hoy
- Forest Gregg, DataMade
- Maryam Judar, Citizen Advocacy Center
- Alaina Harkness, MacArthur Foundation
- Susana Vasquez, LISC
- Terry Mazany, The Chicago Community Trust
- Jean Butzen, Mission + Strategy Consulting
- Emile Cambry, Blue1647
- Laura Sanchez, SWATware, LLC
- Sandee Kastrul, i.c. stars
- Geoff Hing
- Anne Cole, Neighborhood Housing Services of Chicago
- Mike Reilley, DePaul University, The Red Line Project
- Josh Kalov, Kalov Strategies

Conference Documenters

The need for quality, detailed documentation during the Chicago School of Data Days facilitated the creation of Smart Chicago's Documenter Program. Documenters were paid hourly to record video, take notes, and perform other supportive tasks that would capture and preserve the conversations that took place at the Data Days conference. We recruited people using a Wufoo form, making sure to ask about skills and previous experiences.

Things we asked our Documenters during recruitment

- Contact information, including social media
- If they currently worked for a media outlet and, if so, which one
- What they would consider their main type of documentation experience: blogger, videographer, social media, newspaper reporter, television reporter, etc.
- A sample of their documentation work
- A link to their documentation work
- Which conference days they were available for

Kyla Williams, Smart Chicago's Director of Operations, managed the Documenters. She led an introductory phone call for the recruited Documenters, then led a formal pre-conference orientation training session.

List of documenters

- Ali Trumbull
- Carley Mostar
- Gabrielle Lyon
- Genevieve Nielsen
- Hannah Kapp-Klote
- Hope Bertram

- Julie Torkelson
- Justin Freeman
- Leah Lavelle
- Lee Kuhn
- Lindsay Muscato
- Liz Baudler
- Omolola Olateju
- Maalika Banerjee
- Marina Goldshteyn
- Mary C. Piemonte
- Nicole Cipri
- Nourhy Beatriz
- Rose Alfriyie
- Stephen Rynkiewicz
- Suyeon Son
- Terrence Thompson
- Veronica Benson

Smart Chicago also contracted VideoTwist Productions to film the sessions and handle live streaming.

We hired so many people, introducing many of them to our field through the Documenters Program. Later, these Documenters would shine a light on other important Chicago events, topics and debates including the Policy Accountability Task Force and the emerging Array of Things Civic Engagement Project.

Picking the Venue

For the venue, these were our criteria:

- we needed a space that would hold about 300 people
- we needed it in just a couple of months
- Lots of flexible space

The Chicago School of Data Days Conference was held in the Willis Tower. The Willis Tower had unused conference room space which was perfect for us—it provided the space and flexibility needed to accommodate the expected number of people. We had complete control over the setup and configuration of rooms. There were also smaller rooms utilized for "unconference"-like meetings between people.

We found that most attendees stayed in the scheduled sessions rather than taking advantage of the open spaces. This could be because of the conference schedule, attendees' personal preferences, or possibly other factors.

To help our participants navigate the conference space, we provided maps and set up large foam signs directing them to different rooms.

Here is a link to an album of pictures of the space: https://flic.kr/s/aHsk1gKFt9

Catering

Honest Foods provided the catering services for the Chicago School of Data—beverages, two breakfasts, two lunches, and afternoon snacks for 300 people.

Here is what we served on Friday, September 19, 2014:

- Breakfast: Bagels & Lox, Granola & Yogurt, Seasonal Fresh Fruit, Coffee & Tea
- Lunch: Thai Chicken, Teriyaki Salmon, Jasmine Rice, Vegetable Stir Fry, Asian Salad, Szechuan Noodle Salad, Bread Basket, Fresh Fruit

- Afternoon Snacks: Pita with Hummus, Wisconsin Cheese Platter, Crudite, Cookies & Dessert Bars

Conference Structure & Agenda

This conference was not a typical panel-presentation style affair. Our speakers were considered facilitators. Facilitators talked for a short time and then anyone was encouraged to speak, share their experiences, or ask questions. There were no more than 3 sessions happening at one time. You can see the full conference agenda along with accompanying notes and video listed on the Chicago School of Data website.

The conference agenda, like the invited speakers, was decided based on the feedback received from interviews and results from the Chicago School of Data census. Sessions were created and organized around the most popular challenges and topics that arose from the ecosystem scan.

Documentation of the Project

Here are the compiled sources of information about the Chicago School of Data — documentation that was worked on throughout the life of the project and helped us do outreach about the work and our purpose.

We wrote several blog posts about the project on the Smart Chicago Collaborative website:

- The Launch of the Chicago School of Data Project, in which we describe the goals and components of the project
- The Chicago School of Data Interview: An Introduction, in which we describe our approach and methods for interviews
- Chicago School of Data: Interview with Chatham Business Association, in which we provide a partial transcript of our interview with Chatham Business Association

- Chicago School of Data: Interview with Jane Addams Resource Corporation, in which we provide a partial transcript of our interview with the JARC
- Chicago School of Data: Interview with the Illinois Sentencing Advisory Council, in which we provide a partial transcript of our interview with the Illinois Sentencing Advisory Council
- System Used in Chicago School of Data Communications, Organizing and Event Planning, in which we share and describe tools
- Mind the Gap: At the Chicago School of Data Days, in which we describe some takeaways from the Gap session at the conference
- Access Gaps Session at the Chicago School of Data Days, in which we describe some takeaways from the Access session at the conference
- Key Takeaways and Raw Data from the Chicago School of Data Census Form

Partners & media outlets also covered the Chicago School of Data:

- Our project was featured on WBEZ's Tech Shift
- Outside the Loop Radio featured the Chicago School of Data
- The Chicago Community Trust featured the School of Data in its 2014 Annual Report
- Data-Smart City Solutions wrote a piece, "The Chicago School of Data: Building a Framework for Chicago's Data Ecosystem"

We also collected many photos of the Chicago School of Data Days in a Flickr album. We recorded the Chicago School of Data Days sessions, and all videos are up on the Smart Chicago YouTube account.

A breakout room at the Willis Tower, the site selected for the Chicago School of Data Days 2014 *(Photo by Daniel X O'Neil)*

Resources

Thanks to our documenters, the convening had an unprecedented amount of raw, in-the-moment documentation. Based on the work of the documenters, we noted and classified the resources and tools shared by the speakers and participants at the School of Data.

Below is a list of tools, tutorials, and networks, and resources recommended during the Chicago School of Data Days plus a few other recommendations from the Smart Chicago Collaborative. Help us make this guide better— email info@smartchicagocollaborative.org.

Tools

Mapping Tools

- ArcGIS for Desktop http://www.esri.com/software/arcgis/arcgis-for-desktop
- QGIS http://www.qgis.org/en/site/
- Mapbox https://www.mapbox.com
- Leaflet http://leafletjs.com/
- Power Map for Excel
- Map Box/Tile Mill https://www.mapbox.com/tilemill/

Tools to Access & Work with U.S. Census Data

- Census Reporter http://censusreporter.org/
- Census.IRE.org http://census.ire.org/
- NHGIS.org https://www.nhgis.org/
- Social Explorer http://www.socialexplorer.com/
- DataFerrett http://dataferrett.census.gov
- IPUMS https://www.ipums.org/

- AmericanFactFinder http://factfinder.census.gov/
- Stack Exchange re: Census Opendata.stack.exchange.com/questions/tagged/census

Survey Tools

- Wufoo Forms http://www.wufoo.com/
- Survey Monkey https://www.surveymonkey.com/
- Qualtrics https://www.qualtrics.com/
- Google Forms https://www.google.com/forms/about/

Email Campaigns

- MailChimp http://mailchimp.com/

Data Scraping, Cleaning & Analysis

- Google Spreadsheets https://www.google.com/sheets/about/
- Quandl https://www.quandl.com/
- Mr. Data Converter https://shancarter.github.io/mr-data-converter/
- Tableau http://www.tableau.com/
- R http://www.r-project.org/
- Microsoft Access
- Microsoft Excel & Excel Macros
- OpenRefine OpenRefine
- Pentaho http://www.pentaho.com/product/data-integration
- CSVkit https://csvkit.readthedocs.io/en/0.9.0/
- Beautiful Soup https://www.crummy.com/software/BeautifulSoup/
- Web Scraping Tutorial https://www.youtube.com/watch?v=yCcSP3GQhho

Data Visualizations & Infographics

- Datawrapper https://datawrapper.de/
- Infogr.am https://infogr.am/
- Tableau http://www.tableau.com/
- D3 https://d3js.org/
- Easel.ly https://www.easel.ly/

Social Media

- Hootsuite https://hootsuite.com
- Buffer https://buffer.com/
- Twitter http://twitter.com

Web & Tool Development

- GitHub https://github.com/
- Bootstrap http://getbootstrap.com/
- WordPress https://wordpress.com

Customer Relationships Management (CRM)

- Salesforce https://www.salesforce.com

Tools for Journalists (and Other Storytellers)

- Journalist's Toolbox http://www.journaliststoolbox.org/
- Knight Lab tools https://knightlab.northwestern.edu/projects/
- Open journalism https://github.com/showcases/open-journalism
- "How to Setup Your Mac to Develop News Applications Like We Do" http://blog.apps.npr.org/2013/06/06/how-to-set-up-a-developers-environment.html

Chicago Cases

- Flu Shot Finder http://chicagoflushots.org/
- MRelief http://www.mrelief.com/
- TIF illumination project http://www.tifreports.com/
- Was my car towed? http://www.wasmycartowed.com/
- Large Lots https://www.largelots.org/
- Crime in Chicagoland http://crime.chicagotribune.com/
- Chicago Stumbler http://chicagostumbler.tumblr.com/
- Red Line Project http://redlineproject.org/

Community Support

Meetups

- Chicago City Data User Group https://www.meetup.com/ChicagoCityData/
- Chi Hack Night https://chihacknight.org/
- Blue1647 Meetup https://www.meetup.com/BLUE1647-Meetup-Tech-Entrepreneurship-Social-Innovation/
- Chicago R User Group https://www.meetup.com/ChicagoRUG/
- Health Data Liberation Meetup https://www.meetup.com/Health-Data-Liberation/
- The Data Scientist Meetup Group http://www.meetup.com/The-Data-Scientist_Chicago/?gj=ej1b&a=wg2_rdmr

Networks of Organizations and Institutions

- Chicago Benchmarking Collaborative
- Chicago Workforce Funders Alliance
- Connect Chicago
- Chicago Jobs Council
- NTEN http://www.nten.org/

- Hive Learning Network http://www.hivelearningnetworks.org/
- National Neighborhood Indicator Project http://www.neighborhoodindicators.org
- Common Data Set Initiative
- Voluntary System of Accountability
- Digest of Educational Statistics
- National Center for Education Statistics

Some Funders of this Work

- MacArthur Foundation
- Knight Foundation
- Chicago Community Trust
- Boeing Foundation
- Grand Victoria Foundation
- Polk Brothers Foundation

Open Data Sets

Note that this is just a glimpse of the open data available to researchers and workers. We're highlighting the sources cited during the Chicago School of Data Days & Census — some of which were expanded on in this book.

Data Portals & Open Data Maps

- City of Chicago Data Portal
- Cook County Data Portal
- State of Illinois Data Portal
- United States of America Data Portal
- Plenar.io http://plenar.io/
- OpenGrid http://opengrid.io

Education Data

- Association of American Universities Data Exchange http://aaude.org/
- Integrated Postsecondary Education Data System http://nces.ed.gov/ipeds/
- Fountass & Pinnell literacy data http://resources.fountasandpinnell.com/Login.aspx?ReturnUrl=%2fproducts%2fDefault.aspx
- National Center for Education Statistics http://nces.ed.gov/
- Institute of Education Sciences http://ies.ed.gov/
- Illinois State Board of Education http://www.isbe.net/
- Illinois Report Card http://www.illinoisreportcard.com/
- Illinois Longitudinal Data System http://www.ibhe.state.il.us/ILDS/materials/ILDSReport052815.pdf
- National Student Clearinghouse http://www.studentclearinghouse.org/

Resources for Elections and Politics

- Illinois State Board of Elections http://www.elections.il.gov/VotingInformation/ComputerizedVoterData.aspx
- FOIA Machine https://www.foiamachine.org/
- Illinois Sunshine https://www.illinoissunshine.org/

Environmental Data

- Environmental Protection Agency https://developer.epa.gov/category/data/
- Array of Things (coming soon on the Chicago Open Data Portal)

Poverty & Human Services

- SNAP data from Food Research & Action Center http://frac.org/
- Illinois Department of Children and Family Services https://www.illinois.gov/dcfs/Pages/default.aspx
- Social Impact Research Center at the Heartland Alliance https://www.heartlandalliance.org/research/

Housing

- Cook County Assessor http://cookcountyassessor.com/default.aspx?aspxerrorpath=/Property_Search/Property_Search.aspx
- Cook County Recorder of Deeds http://cookrecorder.com/
- Cook County Clerk of the Court http://www.cookcountyclerkofcourt.org/
- Institute for Housing Studies — DePaul University https://www.housingstudies.org/?q=node/3
- Woodstock Institute http://www.woodstockinst.org/
- U.S. Department of Housing & Urban Development http://data.hud.gov/

Justice

- Expungio.io http://www.expunge.io/
- Crime & Punishment http://crime-punishment.smartchicagoapps.org/
- Invisible Institute http://invisible.institute/
- Gang Violence Reduction Project https://www.crimesolutions.gov/ProgramDetails.aspx?ID=278%20
- Sex offender registry https://data.cityofchicago.org/Public-Safety/Sex-Offenders/vc9r-bqvy
- University of Chicago Crime Lab http://crimelab.uchicago.edu/

Labor and Workforce Data

- Longitudinal Employer-Household Data http://lehd.ces.census.gov/data/
- Bureau of Labor Statistics http://www.bls.gov/

Philanthropy Data

- Foundation Center http://foundationcenter.org/
- GuideStar http://datadictionary.cityofchicago.org/
- National Center for Charitable Statistics http:/www.nccs.urban.org/

Additional Reading

The National Neighborhood Indicators Partnership's guide to Data sharing http://www.neighborhoodindicators.org/library/guides/nnip-lessons-local-data-sharing

Data User Guides from WPRDC http://www.wprdc.org/data-user-guides/

City of Chicago's Tech Plan http://techplan.cityofchicago.org/

Data + Design http://orm-atlas2-prod.s3.amazonaws.com/pdf/13a07b19e01a397d8855c0463d52f454.pdf

Leap of Reason https://www.amazon.com/Leap-Reason-Managing-Outcomes-Scarcity-ebook/dp/B0050VHAZM

Creating Value in Nonprofit Business Collaborations http://www.amazon.com/Creating-Value-Nonprofit-Business-Collaborations-Thinking/dp/1118531132

Expanding Training on Data and Technology to Improve Communities from the National Neighborhood Indicators Partnership and Microsoft's Civic Technology Engagement Team http://www.neighborhoodindicators.org/activities/projects/expanding-training-data-and-technology-improve-communities

Acknowledgements

Our last name is "Collaborative." It's how we live and how we work. With a team of just five people, there's really no other way. Here's a quasi-chronological list of the people who've helped us with the Chicago School of Data, and a short description of the huge contributions of each.

Thank you to our Chicago School of Data Days speakers and interviewees for sharing their work and enriching the community of practice.

We also thank our supporters, partners, and community champions:

Daniel X. O'Neil, Smart Chicago's former Executive Director. His leadership and vision brought the Chicago School of Data Days, as well as many other innovative open data and civic technology initiatives, into being. More here: http://www.derivativeworks.com/, here: http://www.danxoneil.com, and here: https://twitter.com/danxoneil.

Julia Stasch, http://www.macfound.org/about/people/170/ Julia was a leader in the formation of Smart Chicago and has articulated many of the precepts of the CUTGroup for years. When writing the foreword to the 2007 report, "The City that Networks," https://www.cityofchicago.org/content/dam/city/depts/doit/supp_info/DEI/CityThatNetworks.pdf, she said, "digital excellence is achieved when all who wish to can integrate the Internet comfortably into their lives—a state of active and meaningful participation that increases knowledge and enhances connections across time and place."

Terry Mazany, http://www.cct.org/about/our-staff/terry-mazany Terry's commitment to inclusion imbues all of our work, as it emanates from the Chicago Community Trust. His tenure at the Trust made its mark on local philanthropy and shaped our work as well as the work of many collaborative and civic engagement initiatives in Chicago.

Brenna Berman, http://www.uilabs.org/team/brenna-berman/ Brenna is the former Chief Information Officer of the City of Chicago and currently leads City Digital at UILabs. She gave a Breakfast/Plenary Session kick-off speech at the Chicago School of Data Days.

Danielle DuMerer, https://www.cityofchicago.org/city/en/depts/doit/auto_generated/doit_leadership.html Danielle DuMerer is the current CIO for the City of Chicago and has been a key partner and champion on many of Smart Chicago's projects.

Kathryn Pettit, http://www.urban.org/author/kathryn-ls-pettit Kathryn is the Senior Researcher and Director of the National Neighborhood Indicators Partnership out of the Urban Institute. She was a plenary session speaker at the Chicago School of Data Days and has been a great supporter and convener of neighborhood data practitioners around the country.

Alaina Harkness: Alaina is a former member of the Operations Committee for Smart Chicago and has always encouraged Smart Chicago to try new things. She is currently a fellow at the Brookings Institute.

Thank you to the Smart Chicago staff, documentaries and consultants, especially:

Kyla Williams: As the second employee of Smart Chicago, Kyla is critical to all of our work. Her years of service in nonprofits and an ease with all people makes her an invaluable voice.

Sonja Marziano: Sonja is the third employee of Smart Chicago and began running the CUTGroup in the Spring of 2014. She developed the system we use for analysis as well as for publishing raw test results. She runs the entire process—meeting with the developers, designing tests, and administering test day like nobody's business.

Adam Garcia: Adam is a long-time Smart Chicago partner and collaborator and played a key role in implementing interview and information gathering for the project.

Andrew Seeder: Andrew is a consultant with Smart Chicago who assisted with writing for the Chicago School of Data Days blog posts and compiling book content.

Lindsay Muscato, http://lindsaymuscato.com/ Lindsay organized, edited, and produced this book and many more of Smart Chicago's publications.

Christopher Whitaker, http://civicwhitaker.com/ Christopher is a leader in the Chicago civic hacker community. His steady output for Smart Chicago was always a huge asset to us and is now compiled in the @CivicWhitaker Anthology https://www.amazon.com/CivicWhitaker-Anthology-organizing-documenting-Collaborative/dp/0990775259

Josh Kalov: Josh Kalov of Kalov Strategies is a consultant for Smart Chicago and leads our Cook County Open Data project. Josh also advises Smart Chicago and several of our projects. He was a major contributor to the Chicago School of Data conference. As a longtime member of the open government community in Chicago, we value his input on nearly every topic.

Matt Gee: Matt led the Chicago School of Data organizational scan in its early stages, collecting notes, knowledge, and insights that would become the Chicago School of Data Days.

Most of all, thank you to the more than 250 Chicago organizations who responded to the School of Data Census and participated in the School of Data Days. You're it.

The Editor

Denise Linn Riedl is the Program Analyst for the Smart Chicago Collaborative. She joined the team in 2015 and manages collaborative digital inclusion and smart city initiatives like Connect Chicago, and the Array of Things Civic Engagement Project. She is a champion for abundant, affordable broadband and inclusive, resident-informed technology projects. In 2015, she published "A Data-Driven Digital Inclusion Strategy for Gigabit Cities" and co-wrote the "Next Generation Network Connectivity Handbook." She previously worked at the Federal Communications Commission and is also an AmeriCorps VISTA alumna. Denise has a bachelor's degree in Media Studies and English from the University of Virginia and master's in public policy from the Harvard Kennedy School of Government. You can follow her on Twitter here: https://twitter.com/dklinn

Made in the USA
Monee, IL
04 November 2026

67384432R00066